购物中心

香港视界国际出版有限公司 主编

華中科技大學出版社
http://www.hustp.com
中国·武汉

## 图书在版编目（CIP）数据

购物中心 / 香港视界国际出版有限公司 主编 . – 武汉 : 华中科技大学出版社 , 2011.10

ISBN 978-7-5609-7279-4

Ⅰ . ①购··· Ⅱ . ①香··· Ⅲ . ①购物中心 – 建筑设计 – 图集 Ⅳ . ① TU247.2-64

中国版本图书馆 CIP 数据核字（2011）第 166825 号

购物中心　　香港视界国际出版有限公司 主编

出版发行：华中科技大学出版社（中国・武汉）
地　　址：武汉市武昌珞喻路1037号（邮编430074）
出 版 人：阮海洪

责任编辑：熊纯　　责任监印：张贵君
责任校对：彭江　　制　　作：百彤文化

印　　刷：利丰雅高印刷（深圳）有限公司
开　　本：889 mm × 1194 mm　1/12
印　　张：22
字　　数：132千字
版　　次：2011年10月第1版 第1次印刷
定　　价：298.00元（US $46.99）

投稿热线：（010）64155588-8000　hzjztg@163.com
本书若有印装质量问题，请向出版社营销中心调换
全国免费服务热线：400-6679-118 竭诚为你服务

# PREFACE

# 前言

## PREFACE DRAFT
## EXPERIENCE + INNOVATION
### Retail Design – Planning for Success
### By John Simones, Partner, Director of Design, The Jerde Partnership, Inc.

By John Simones, Partner, Director of Design, The Jerde Partnership, Inc.

No matter how quickly technology continues to evolve and how many new gadgets or social media devices are developed, people will continue to desire communal experiences – the personal exchange of being with someone else; being in a group; seeing people and being seen by people. As the world inevitably becomes more virtual, people will also demand more visceral environments. In order to accomplish this, we focus on the building type that is the only remaining vestige of a communal scene – the Shopping Center.

Places that provide various offerings to purchase, and events that entertain – retail / leisure projects – are the places that will satisfy people's need to be together. Our goal is to focus on the public realm associated with retail development in order to transform the shopping center into an urban village that becomes the communal soul of a host City – whether it is Shanghai, Los Angeles, New York, Rotterdam, Warsaw, Tokyo, or Seoul – people want to interact and be social, but they need activated and compelling environments to do so.

How should these places be designed and what should happen there?

The success of our retail design work has been our ability to solve issues through what we call the "The Big Idea" – a design solution that creates an experience and place that is so compelling it will attract people in large numbers, provide pleasant memories so customers return with their family and friends, and cause huge economic success. These solutions include a spatial quality that is inspired by the local context and aspirations of the local citizens to create an authentic and vibrant project that becomes the most popular destination in the region. The spaces within the project are interwoven with the surrounding pedestrian circulation and are highlighted with contextual textures and colors along with a landscape that attracts people of all ages. Likewise the building shapes are formed to reinforce the spaces within and to establish an exterior identity and landmark within the city. This approach is both practical and artistic. The solution, despite its innovative qualities, feels as if the project belongs to the city thus grows in popularity over time, delivering a timeless quality.

There are two basic issues that make a retail project successful.

First, it must be a place that inspires people to engage in the human experience, and establish a sense of belonging. The project should not feel like a mall, but rather an extension of the city, thus becoming the most popular place to go by appealing to people's enjoyable emotions. By creating zones, or districts, within a single place, each expressing qualities of time, space, light, form, and texture, combined with issues of use, mood, vitality, geography, and human reactions to place, we develop retail and mixed-use projects based on discovery and experiential design. Therefore, the architecture and space is not a predetermined solution or style, rather a customization that is conceived to react and mature with the ever-changing demands of people's likes and dislikes, which ultimately creates a project that will attract increased numbers of people.

Second, the offering of various goods, food, beverages, entertainment and services must be configured in a way that maximizes the exposure for utilization, while creating new nodes of social interaction. Most competing retail projects offer essentially the same tenant mix. However, if that offering is situated within an attractive environment and arranged along a well thought out organizational framework and path of discovery, the space and environment will feel unique to the user. The planning and design of successful retail projects requires a strong understanding of the behavioral characteristics of consumers. A project must appeal to a collection of audiences: adults and families; children; young adults and professionals; seniors; locals; tourists; and the like. By understanding the desires of consumers, particular attention must be given to tailoring specific environments that cause an enjoyable emotional experience for each consumer group. Incorporated into a Jerde design are places that express the individual spirit of the locale, places for festive events, daily markets, and community gatherings. As a collective whole, we call this "heartmaking," which by delivering memorable experiences and spaces for people will drive enhanced economic success for the project.

In order to create these spaces, we focus on the need and desire of people wanting an experience while shopping. It is generally acknowledged by leading retailers that enhanced social destinations tend to attract people longer, leading to additional purchases. Therefore, creating a communal place that people return to will drive retail sales much higher than creating a standard, uninspiring solution.

For a retail project to be truly successful, the design must acknowledge, accommodate and encourage the activity of shopping. Historically the exchange of goods took place in a marketplace, along the downtown "main street," in a bazaar with street vendors, or in a farmer's market. These were not typical mall environments; they were a part of everyday activity and created over time in response to the characteristics of the City and people that lived there. These are the true inspirations for our design of retail spaces of today.

无论技术持续发展的速度有多快，无论开发了多少新玩意或社会媒体设备，人们仍然渴望集体经历——跟其他人在一起的人际交流，在一个集体里，看到人与被人看到。由于世界不可避免地变得更加实际，人们也将要求更多的发自内心的环境。为了实现这一点，我们专注于购物中心这种建筑类型，它是公共场景唯一剩下的遗迹。

人们可以购买各种东西，或者进行娱乐活动(零售或休闲项目)的地方将是可以满足人们聚在一起的这种需求的好去处。我们的目标是专注于与零售业的发展相关的公共领域，以便把购物中心转化为城中村，成为主办城市的公共灵魂。无论是上海、洛杉矶、纽约、鹿特丹、华沙、东京或首尔，人们想要互动和社交，但是他们同样需要活跃和令人信服的环境。

这些地方应该如何设计，应该做些什么?

我们的零售设计工作的成功是通过我们所说的“大创意”来提升解决问题的能力，“大创意”是一个能创建激发兴趣的经历和地方的设计解决方案，它会吸引大量的人，并为顾客提供愉快的回忆，所以顾客会带着他们的家人和朋友回到这里，并创造巨大的经济效益。这些解决方案包括空间的品质，它受到当地背景和当地市民愿望的激励，创造出一个真实并充满活力的项目，并且该项目成为了该地区最热门的目的地。该项目里面的空间和周围的行人流通交织在一起，并因背景的纹理，颜色以及一条吸引了各种年龄层次的顾客的风景线而成为突出的亮点。同样地，建筑形状的形成是为了加强里面的空间，并在全市范围内建立一个外部特征和地标。这种方法既实用又有艺术性，不论其创新的特质，这解决方案让人感觉好像该项目是属于这个城市的，并随着时间推移而大受欢迎，且传达着一种永恒的特质。

有两个基本问题，解决了，零售项目即可成功。

首先，它必须是一个可鼓励人们参与人类体验和建立一种归属感的地方。项目不应感觉像个商场，而是城市的一个延伸，由于使人们心情愉悦而成为最受欢迎的地方。通过在一个独立的空间建立地带或地区，使每个地带和区域都体现时间、空间、光、形状和纹理的特性，结合使用、情绪、活力、地形和人们反应的问题，我们是基于发现与经验设计而发展零售和混合使用的项目。因此，建筑空间不是一种预定的解决方案或样式，而是一种定制，设想对人们喜恶不断变化的要求而做出反应并使之成熟，最后创建一个项目可吸引越来越多的人。

其次，各种各样的商品、食物、饮料、娱乐和服务的提供必须以最大限度地使用方式进行配置，同时创造社会互动的新节点。大多数竞争性的零售项目本质上提供相同的租户组合。然而，如果提供的配置是位于一个引人入胜的环境里，并且安排在一个经过深思熟虑的组织框架和发现路径中，空间和环境对于用户来说是独一无二的。成功零售项目的规划和设计要求对消费者的行为特征有深刻的理解。一个项目必须吸引到一批观众：家庭、儿童、职业人士、年长者、当地人、游客等等。通过了解消费者的需求，必须特别注意一些量身定制的特定环境，这种环境会为每个消费群体带来一种愉快的情绪体验。纳入Jerde设计的地方有表达场所、举行节日活动地方、日常集市和社区聚会的独特精神。作为一个共同整体，我们把这叫做“用心制作”，通过为人们提供难忘的体验和空间，它将为此项目带来经济方面的成功。

为了建造这些空间，我们着重于人们购物时理想体验的需要和渴望。增强的社会目的往往吸引人们的购物时间更长，从而引起更多的购买行动，这是领先的零售商所公认的。因此，建造一个人们会回来购物的公共地方将会比建造一个普通的、不能引发兴趣的地方更能提高零售销售额。

零售项目要真正成功，其设计必须认可、容纳和鼓励购物的活跃性。历史上商品交换是发生在市场里，在沿着市中心的主要街道，在街头摊贩的集市里或者是在农贸市场里。这些都不是典型的商场环境；他们是日常活动的一部分，并随着时间的推移而创建，以反应城市和生活在那里的人们的特点。这些是我们今天设计零售空间的真正灵感。

目录

# CONTENTS

# Forum Istanbul Shopping Center
# 伊斯坦布尔Forum购物中心

Forum Istanbul Project was developed in a design partnership between ERA, BDP, T+T, and Chapman Taylor for Bayrampasa district where E5 highway marks the south boundary of the site. The east and north ends are lined off by tramlines. The project is a mixed - use development including retail, residential, office and entertainment areas settled in three combined parcels. The project's retail functions cover a shopping center with 100,000 sq.m. leasable area, 15,000 sq.m. hypermarket, 30,000 sq.m. Ikea Store, 10,000 sq.m. DIY stores, and 5,000 sq.m. sports store with large parking space for 5,500 vehicles. 20,000 sq.m. of office area also supports retail functions. The shopping center is planned on four storeys with shops, department stores, restaurants, leisure area, cinemas, and a 7,500 sq.m. aquarium which was designed in the basement levels where ocean life can be discovered. The building is inspired by "city within the city" concept. A traditional city structure has been the inspiration for the design master plan. A city like image and scale is aimed through a combination of different buildings on the exterior as well as interior spaces...

**Location**
ISTANBUL, TURKEY

**Area**
164,000 sq.m.

**Company**
ERA PLANNING ARCHITECTURE CONSULTING CO. LTD.

**Designer**
ERA PLANNING ARCHITECTURE CONSULTING CO. LTD.

伊斯坦布尔Forum项目是ERA,BDP, T+T, 和 Chapman Taylor在Bayrampasa区联合开发的一个项目，该地区的E5高速公路是该地区南部边界的标志，东部和北部两端由电车轨道一字排开。该项目是一个综合用途发展项目，包括位于3个相连区域的零售、住宅、办公和娱乐区。该项目的零售功能涵盖可供出租的10万平方米的购物中心，15,000平方米的大卖场，30,000平方米的宜家商场，10,000平方米的自助商店，以及带有可停5,500辆汽车的大型停车场和5,000平方米的体育商店。 20,000平方米的办公面积同时也有零售功能。购物中心规划为四层楼，有商店、百货公司、餐厅、休闲区、电影院和一个7,500平方米的水族馆，这个水族馆以前是设计在能发现海洋生物的是地下室层。该建筑的灵感来自于“城市中的城市”这一概念。一个传统的城市结构为设计总体计划提供了灵感。城市的形象和规模是通过不同的建筑外观以及内部空间的组合而达成的。

U.S. POLO ASSN.
SINCE 1890
TWIST
POLOGARAGE
POLOGARAGE
BLACK LABEL

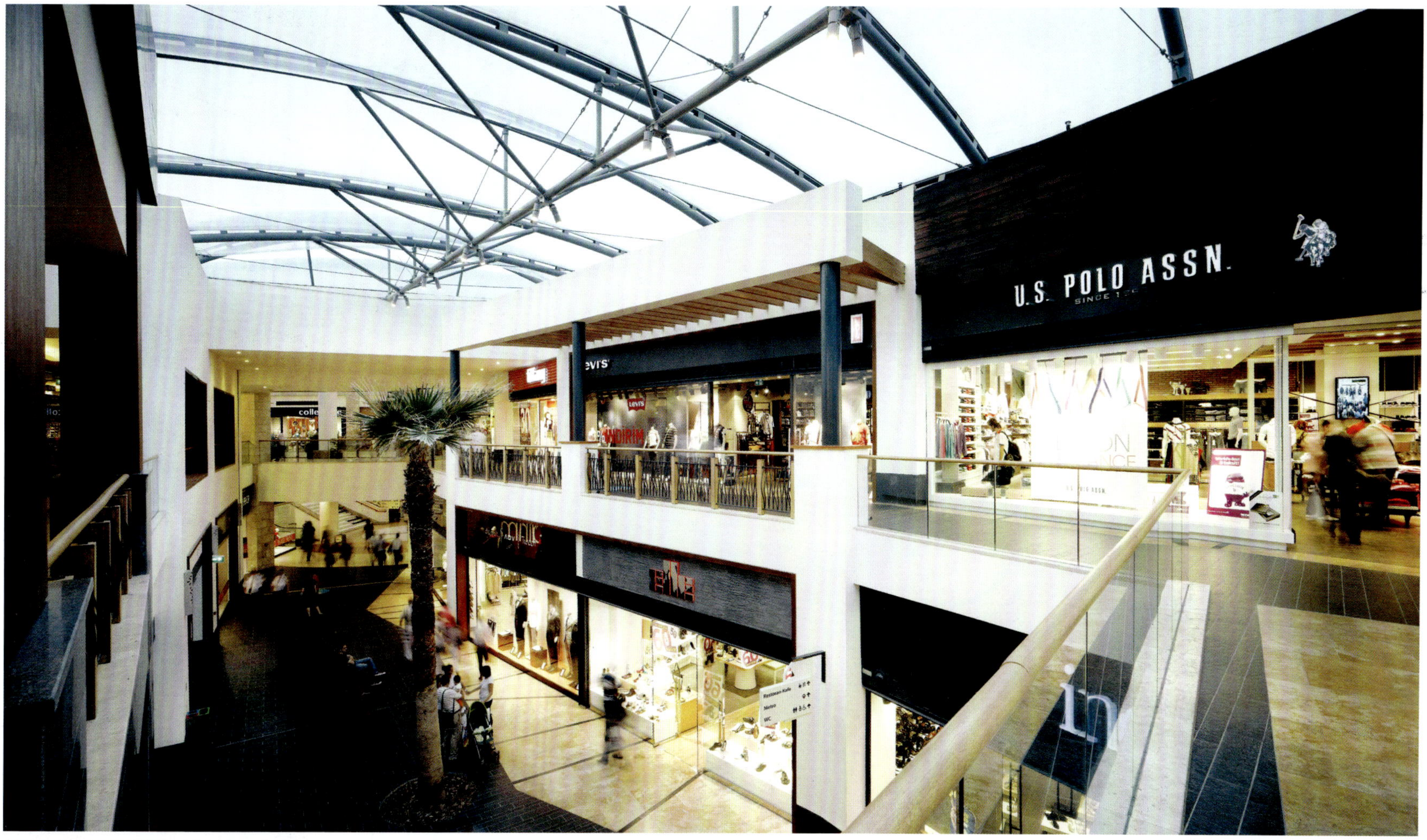
U.S. POLO ASSN.
levi's

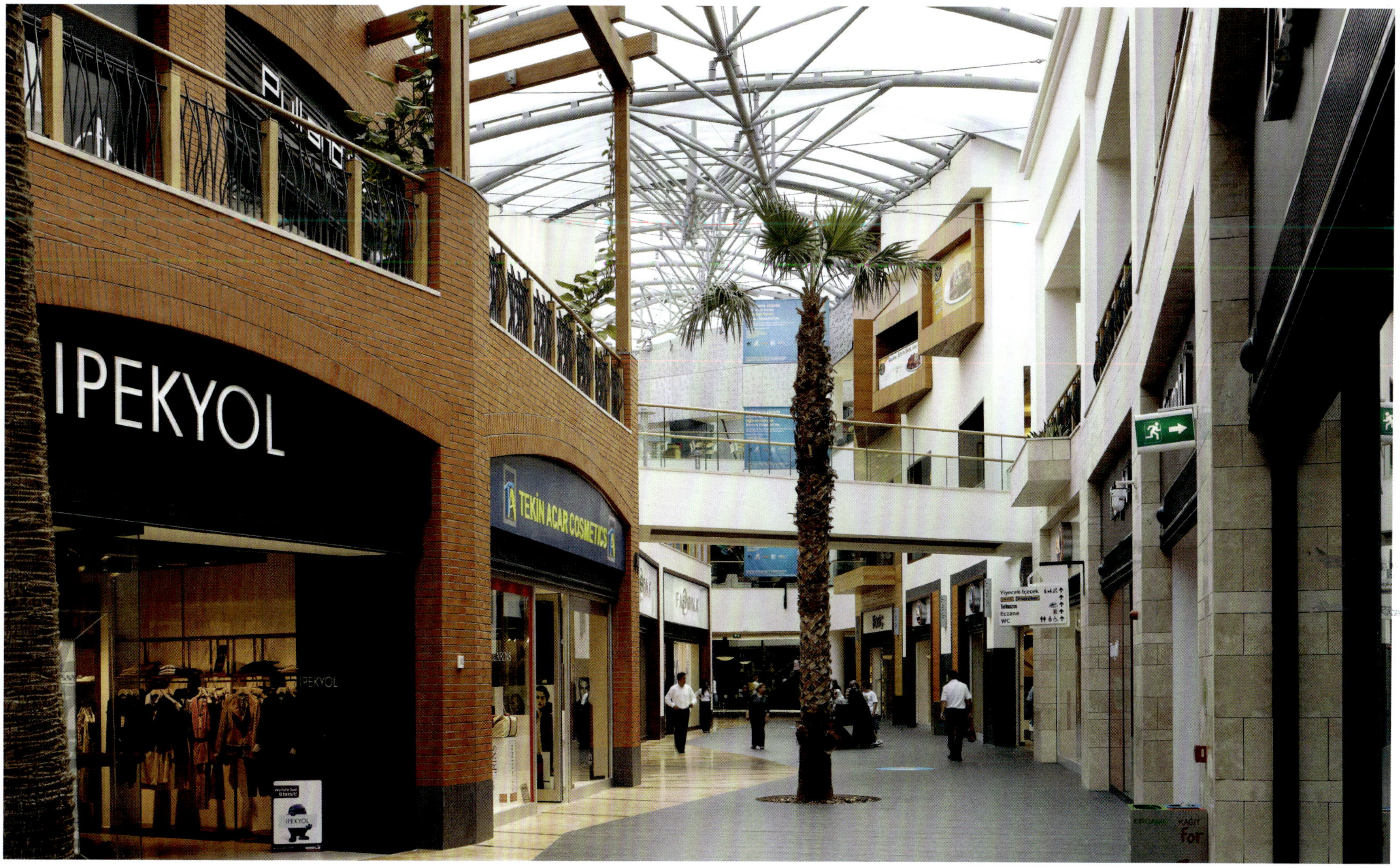
IPEKYOL
TEKİN ACAR COSMETICS

D&R
MUSIC BOOK
STORE
Turkuazoo
AKVARYUM
AKVARYUM

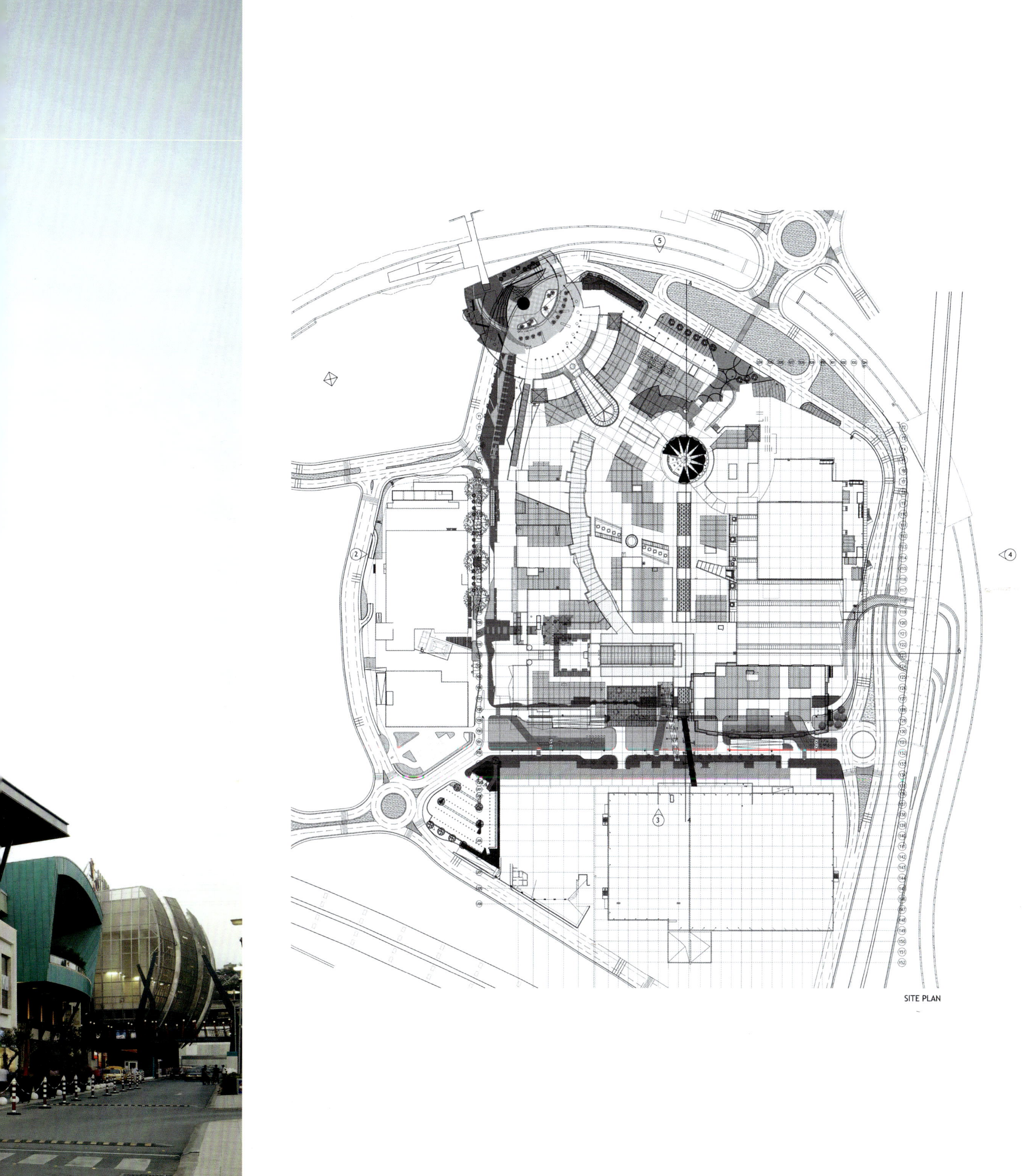

SITE PLAN

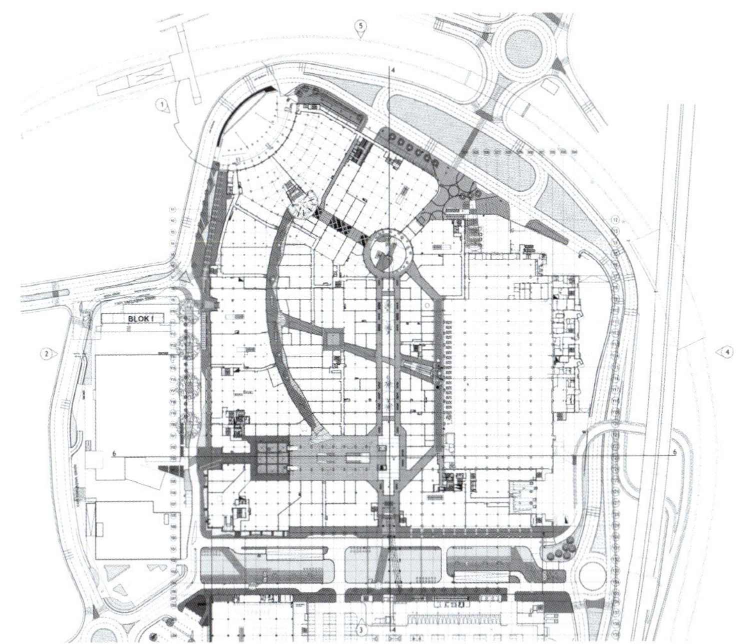

GROUND FLOOR PLAN

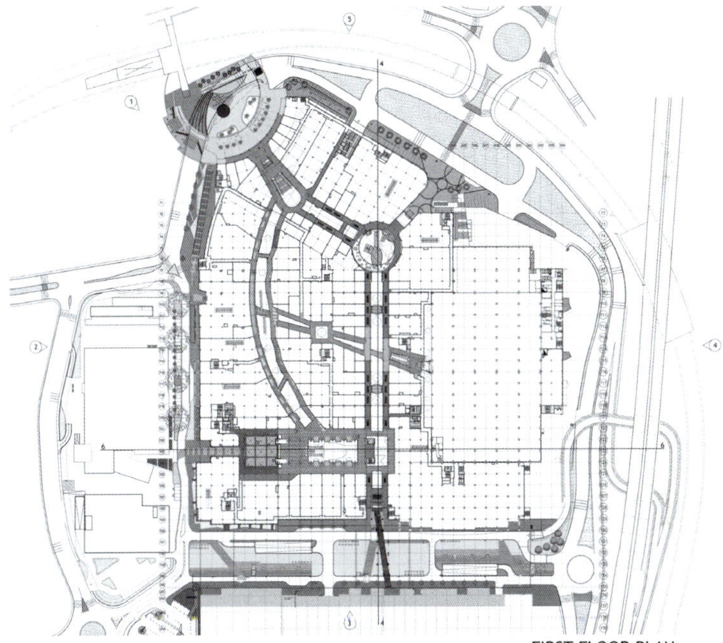

FIRST FLOOR PLAN

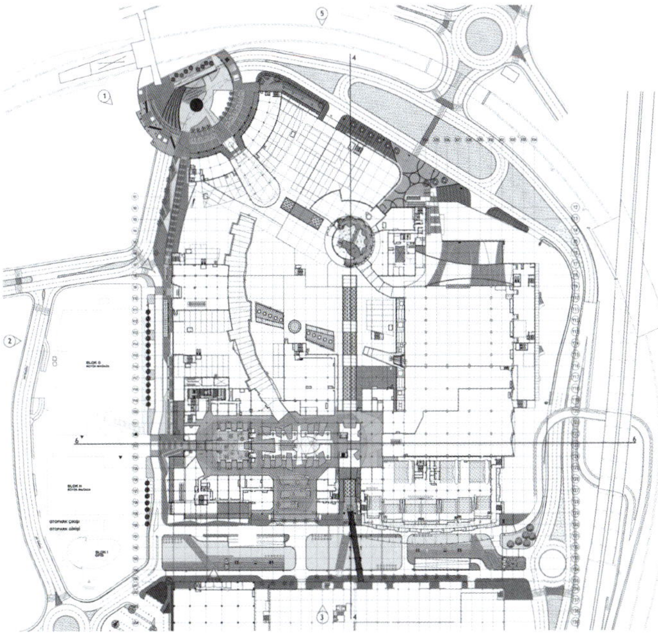

SECOND FLOOR PLAN

VETRA

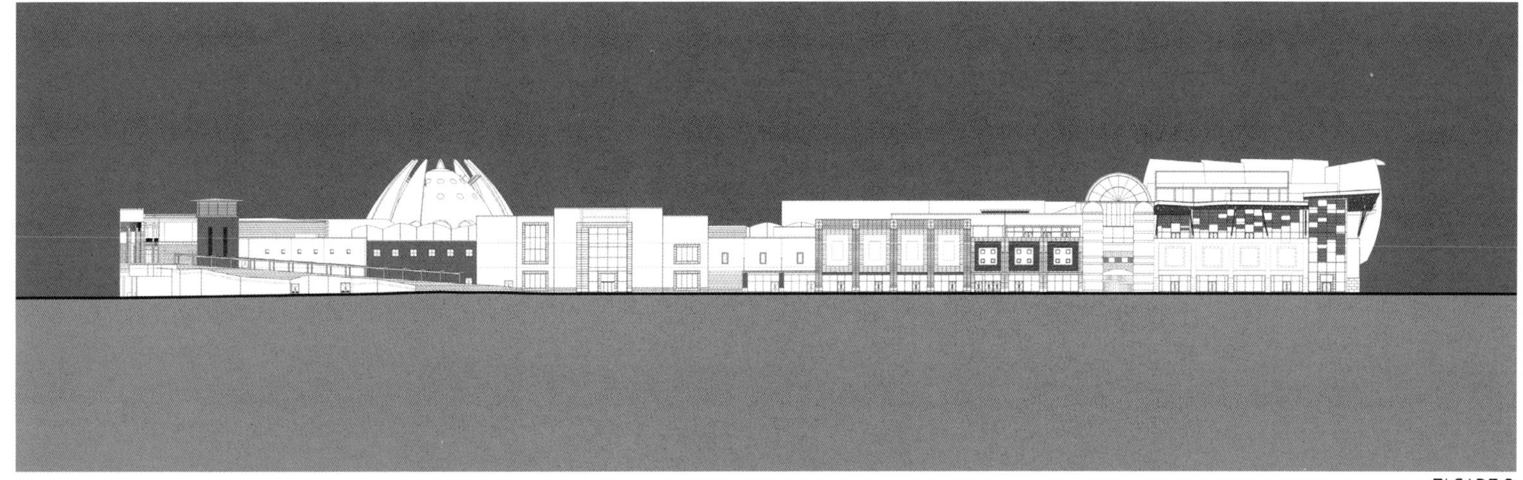
FACADE 2

FACADE 3

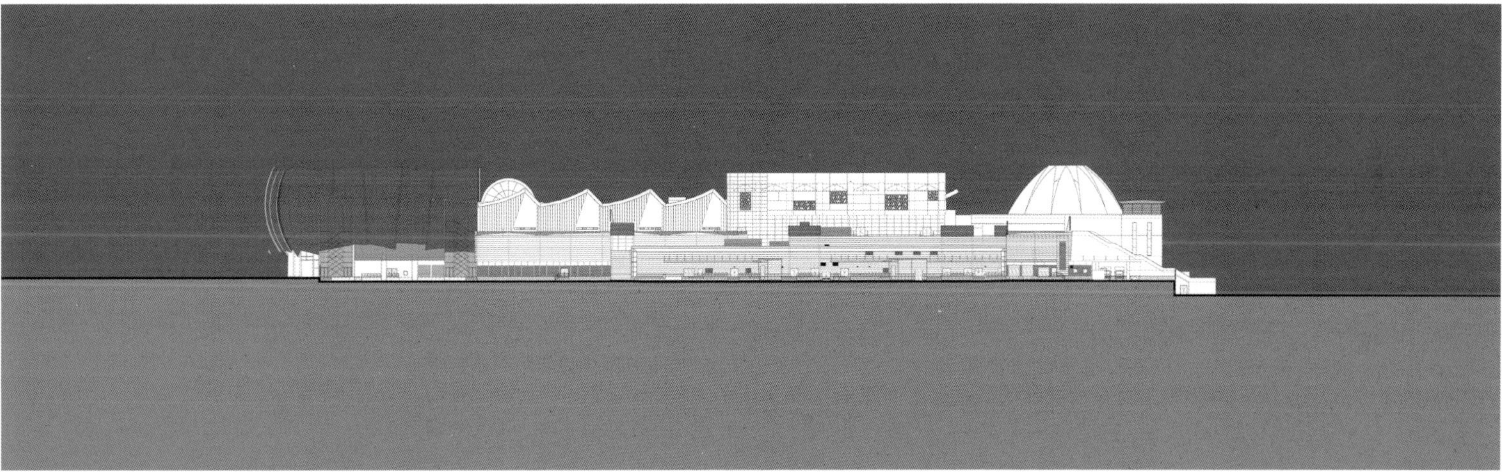
FACADE 4

SECTION

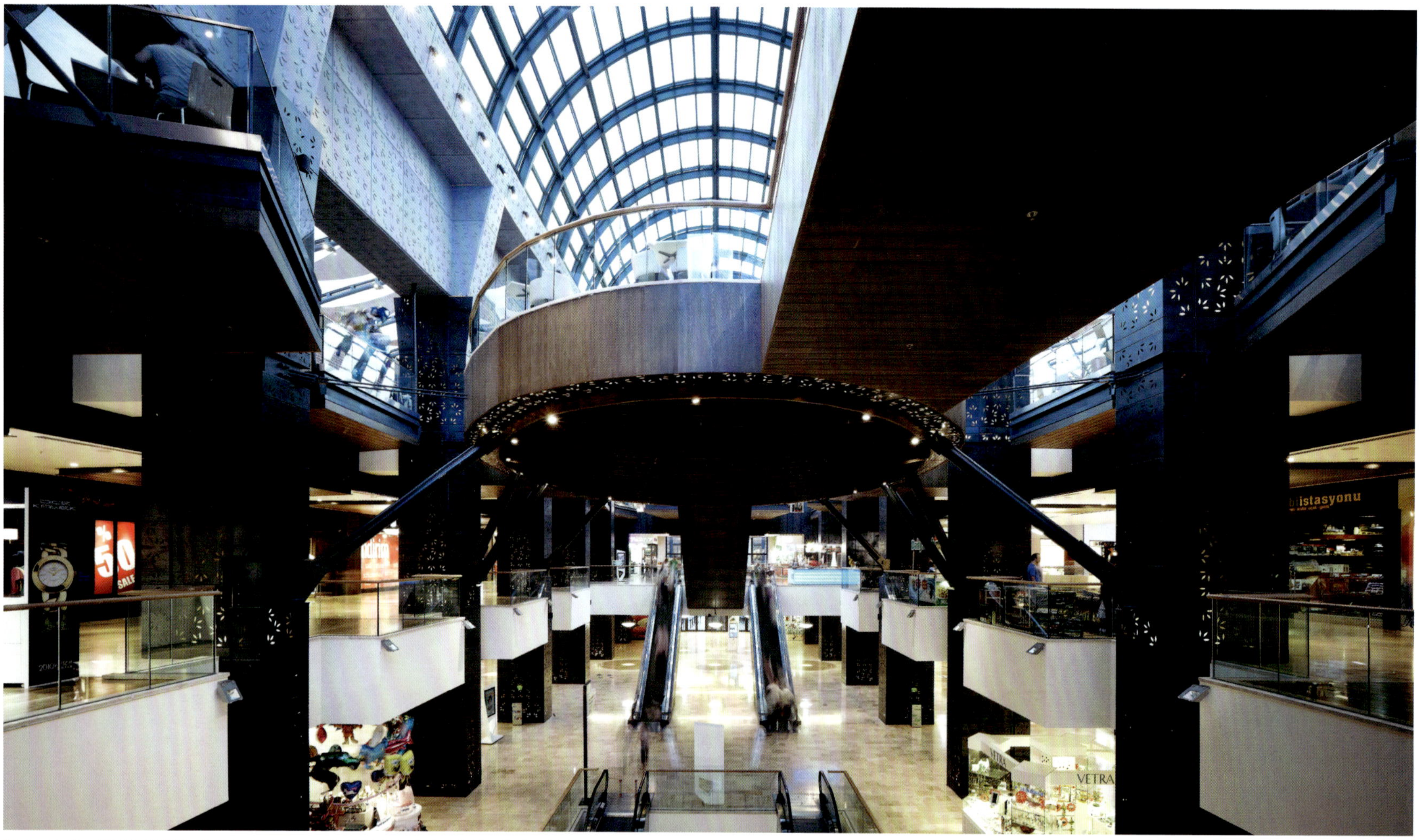
50
SALE
VETRA

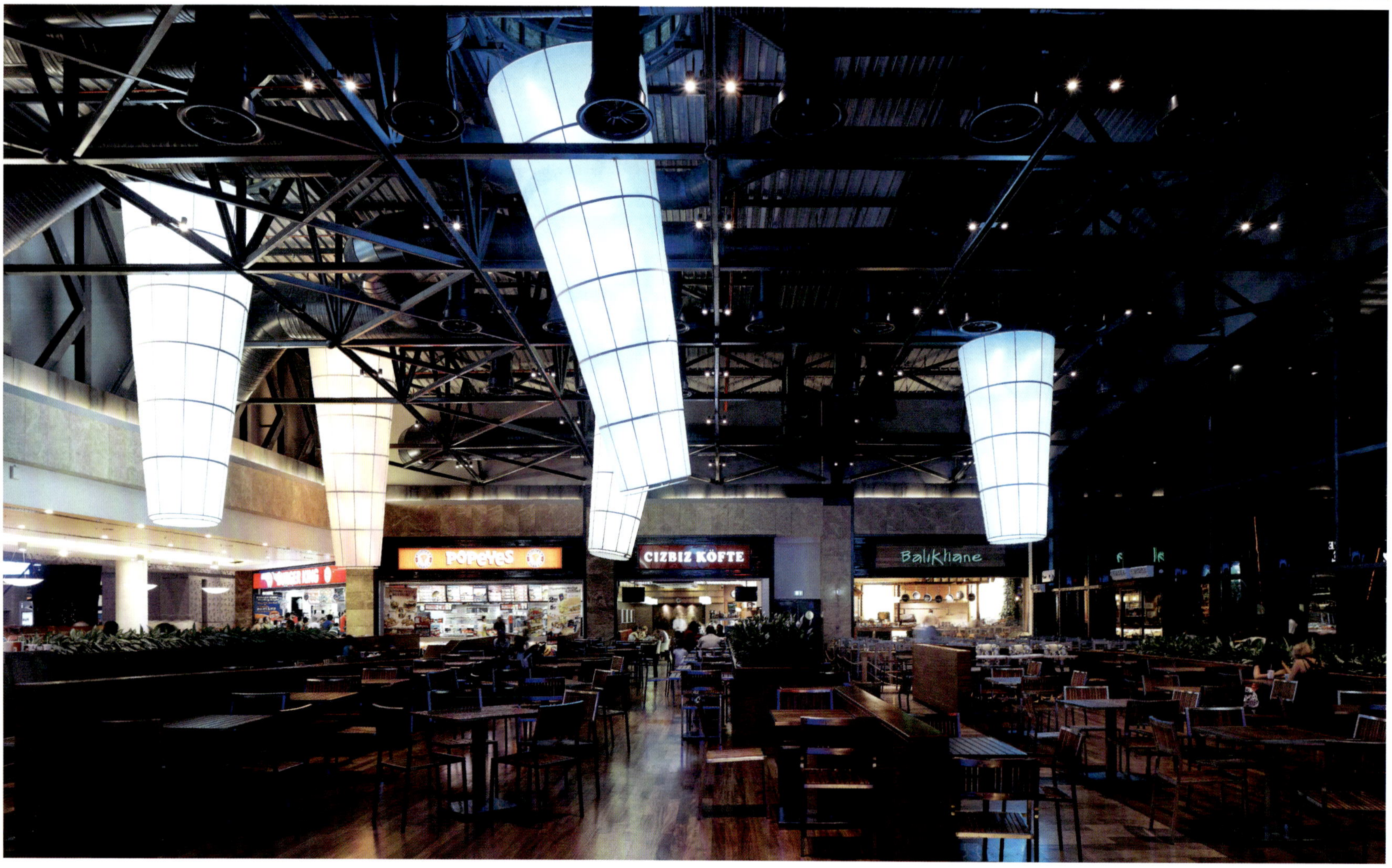
POPEYES
CIZBIZ KÖFTE
Balıkhane

# 360 MALL

## 360购物中心

360 Mall was envisioned as a hub of retail, entertainment and leisure offerings that would act as an indoor civic center and an extension of everyday living where residents and tourists can gather to socialize and shop.

The name 360 guides all elements of the scheme and references organic and manmade symbols, from the rotation of the Earth to the needle on a compass, an important reference to the historic Arabic art of navigation. The distinctive rounded exterior comprises a carved limestone façade surrounded by lush landscaping and water features. Inside, visitors progress through a main entry and into a techno hub, a versatile space designed to showcase information about the center and its sponsors. The retail diagram unfolds in two opposing concourses, "day journey" and "night journey." The design elements in each concourse - from lighting to graphics to interior architecture - respond to the qualities implied by "day" and "night". At the intersection of the corridors, a three - level crescent-shaped atrium features bronze screens that reinterpret traditional Arabic patterns.

Other high-quality materials and plush furnishings allow the atrium to function as event and gathering space. In addition to international and specialty retail tenants, a hypermarket, a cinema and a bowling alley, the center also houses a food court. Diverse tenants and seating configurations depart from the typical line - of - stalls format and offer varied experiences, from quick - service to white tablecloth dining.

**Location**
Kuwait City, Kuwait

**Area**
130,064 sq.m.

**Design Company**
RTKL Associates Inc.

**Photographer**
Mitch Duncan

360购物中心作为一个零售、娱乐休闲中心，将成为一个室内的市民中心，同时也是居民和游客聚集起来进行社交和购物的日常生活的延伸。

360这个名字引导了本方案的所有元素，并参考了一些有机和人造的符号，从地球自转到罗盘针，是历史性阿拉伯航行艺术的一个重大参考。独特的圆形外观，包括一个雕刻了的石灰石外墙，四周被郁郁葱葱的园林绿化和水景环绕着。在里面，参观者通过主入口进入一个高技术音乐中心，这是一个用来展示中心和其赞助商的信息的多功能空间。零售图呈现在两个相对的大厅里，即“白天之旅”和“夜间之旅”。从照明到图表到室内建筑，每个大厅的设计元素都响应着“白天”和“夜晚”所隐含的特性。在走廊的交叉点，3层的新月形中庭以青铜色的屏幕为特色，重新诠释着传统的阿拉伯图案。

其他高级的材料和豪华的家具使中庭成为一个活动和聚会的空间，除了国际和特色零售租户、特大百货市场、电影院和保龄球场，这个购物中心还设有一个美食天地，不同的租户和座位配置跟典型的线型摊位设计不同，并从快餐到铺有白色桌布的饮餐，提供了各种各样的体验。

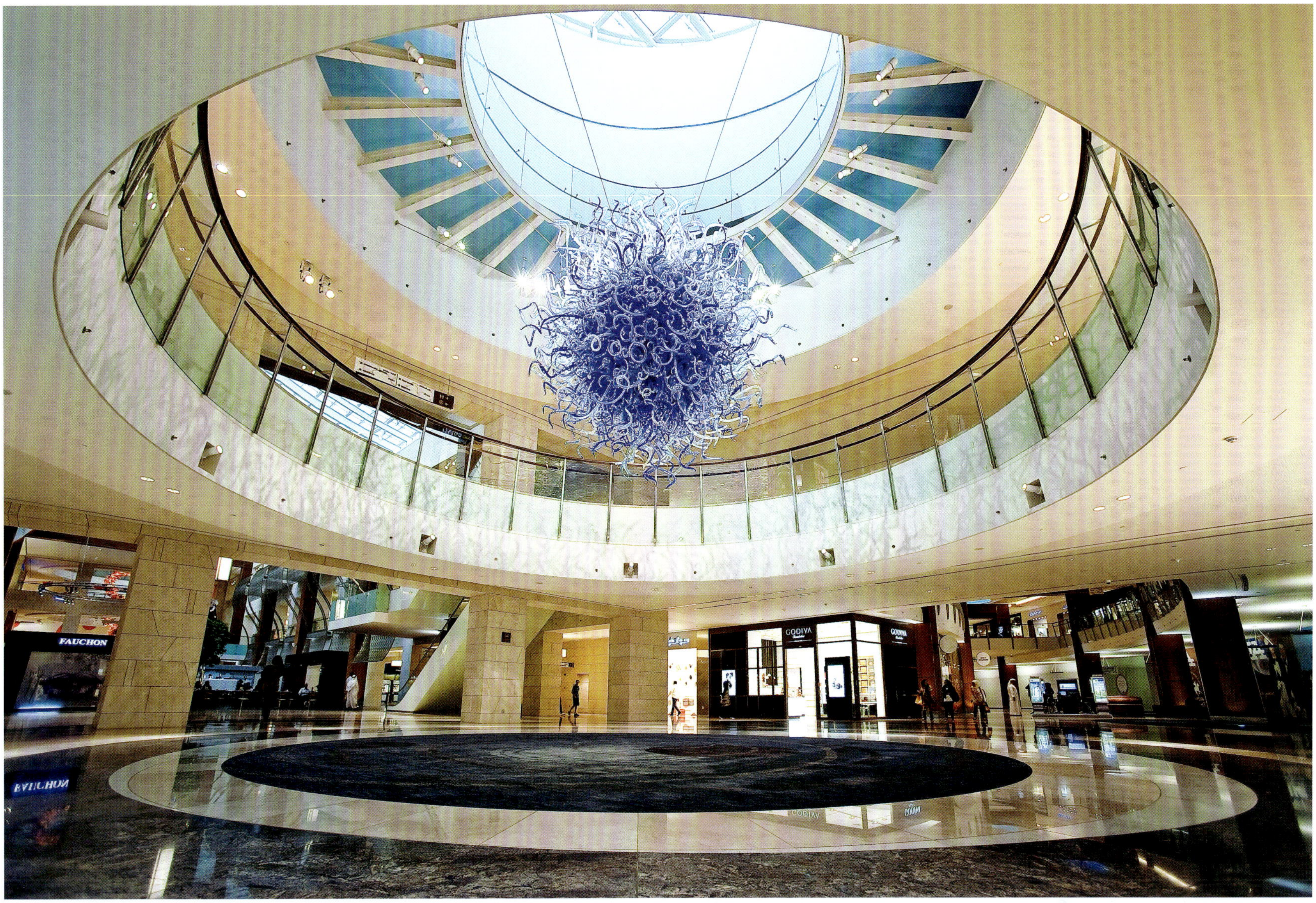
FAUCHON
GODIVA
GODIVA

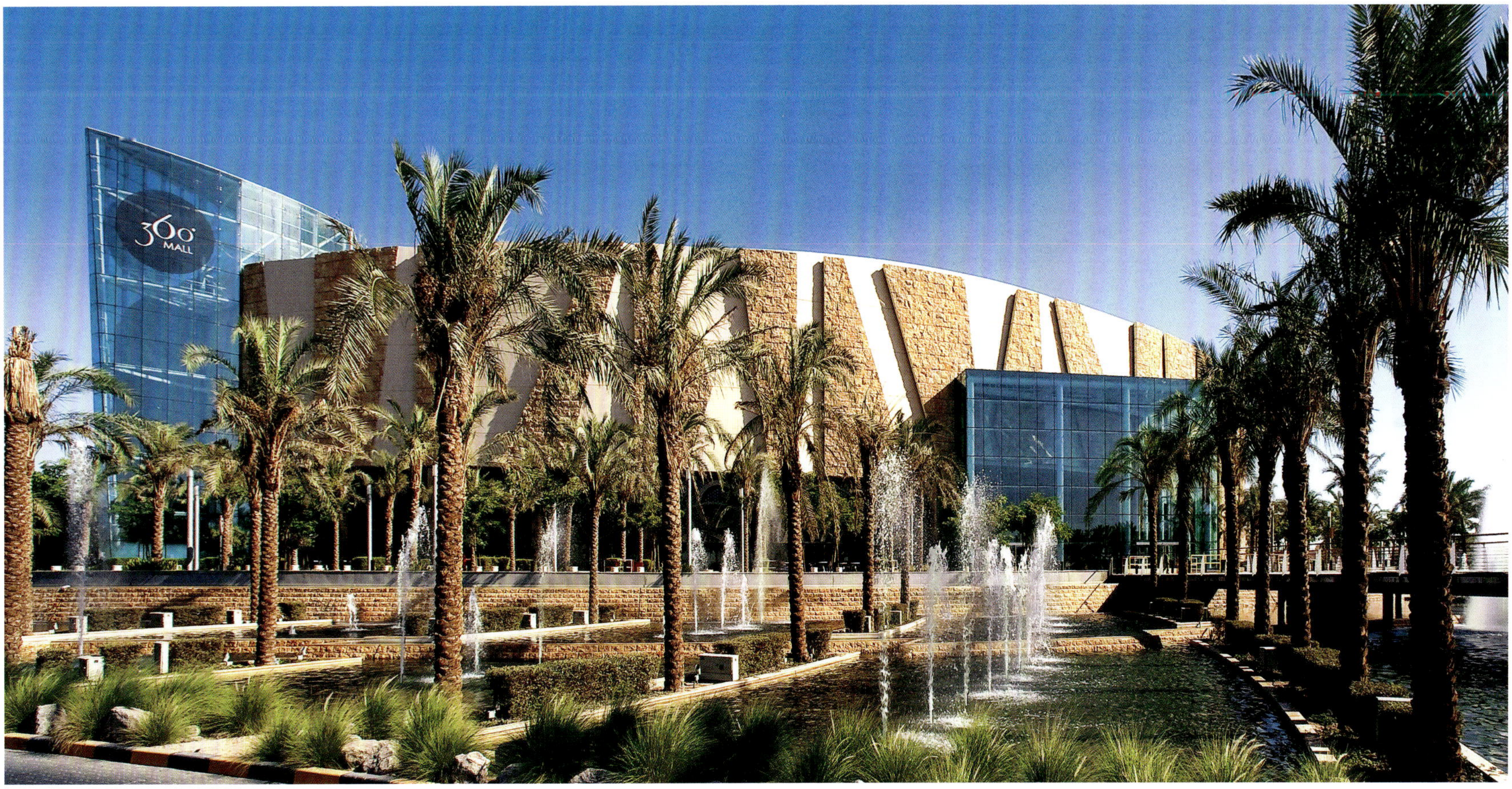
360°
MALL

360°
MALL
Koton

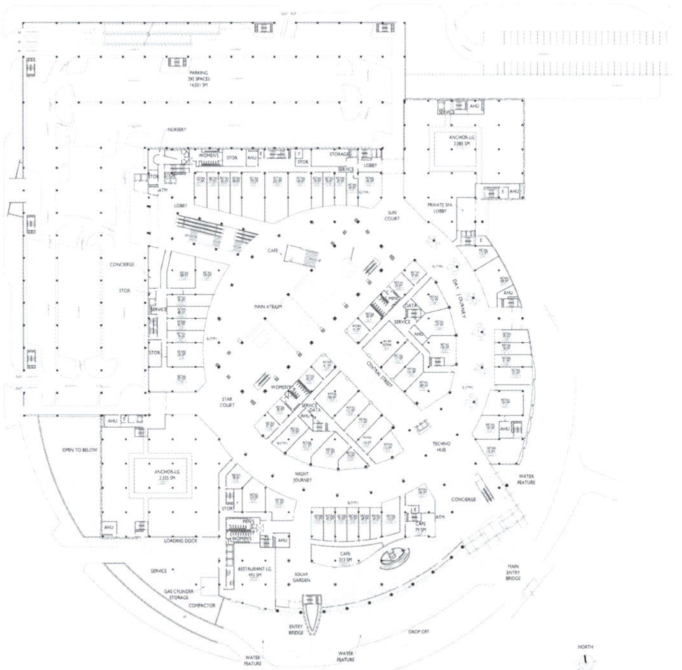

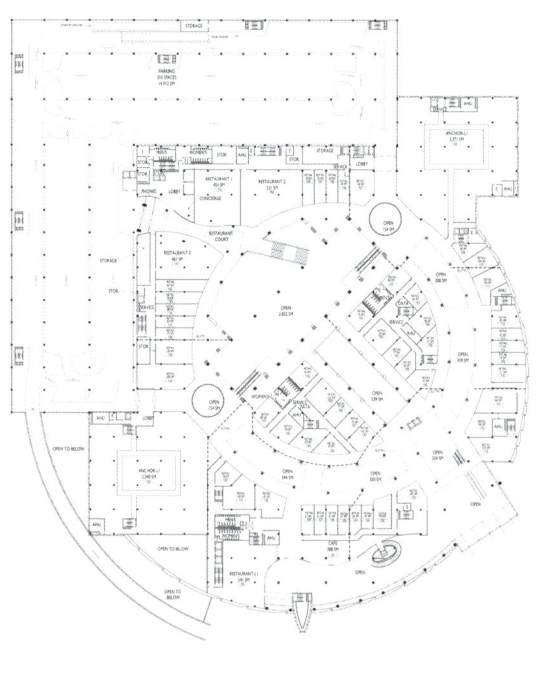

MONTALE

# City Crossing
# 华润中心

The 195,400sq.m. south parcel continues and extends the commercial district southward, here a number of uses including retail, residential, and hotel/hospitality are organized into an "urban neighborhood". The 25,000sq.m. of street-oriented retail extends southward along Boa An Road, a major artery through the city. Here the shopping center shifts from one that is indoor and enclosed to one that is outdoor and open. An elevated bridge connecting the north and middle parcel to the south parcel provides an easy transition to the dynamic development. Three residential towers frame the western edge of the parcel. Surrounded by beautiful landscaped courtyards, parks, water features, and pedestrian friendly streetscapes, the towers are positioned to provide breathtaking views of the Bu Ji River and southeastern portions of the city.

195,400m$^2$的南边地块向南继续延伸该商业区，这里有许多使用区域包括零售、住宅、酒店，组织成一个“城市居民区”。25,000 m$^2$的面向街头的零售店沿着通向城市的主动脉宝儿安路向南延伸。这个的购物中心的转变是从一个室内的封闭的空间到一个室外的开放的空间的改造。高架桥梁将北部与中部的地块和南部的地块连接起来为动态发展提供了一个轻松的过渡。3个住宅塔将此地块的西部边缘围起来。四周被风景优美的庭院、公园、水景和步行区友好的街景围绕，这些塔被定位来提供布吉河和城市的东南部分壮丽景色。

**Location**
Shenzhen, China

**Area**
1,953,934.12 sq.m.

**Company**
RTKL Associates Inc.

**Photographer**
Aaron Pocock and David Whitcomb

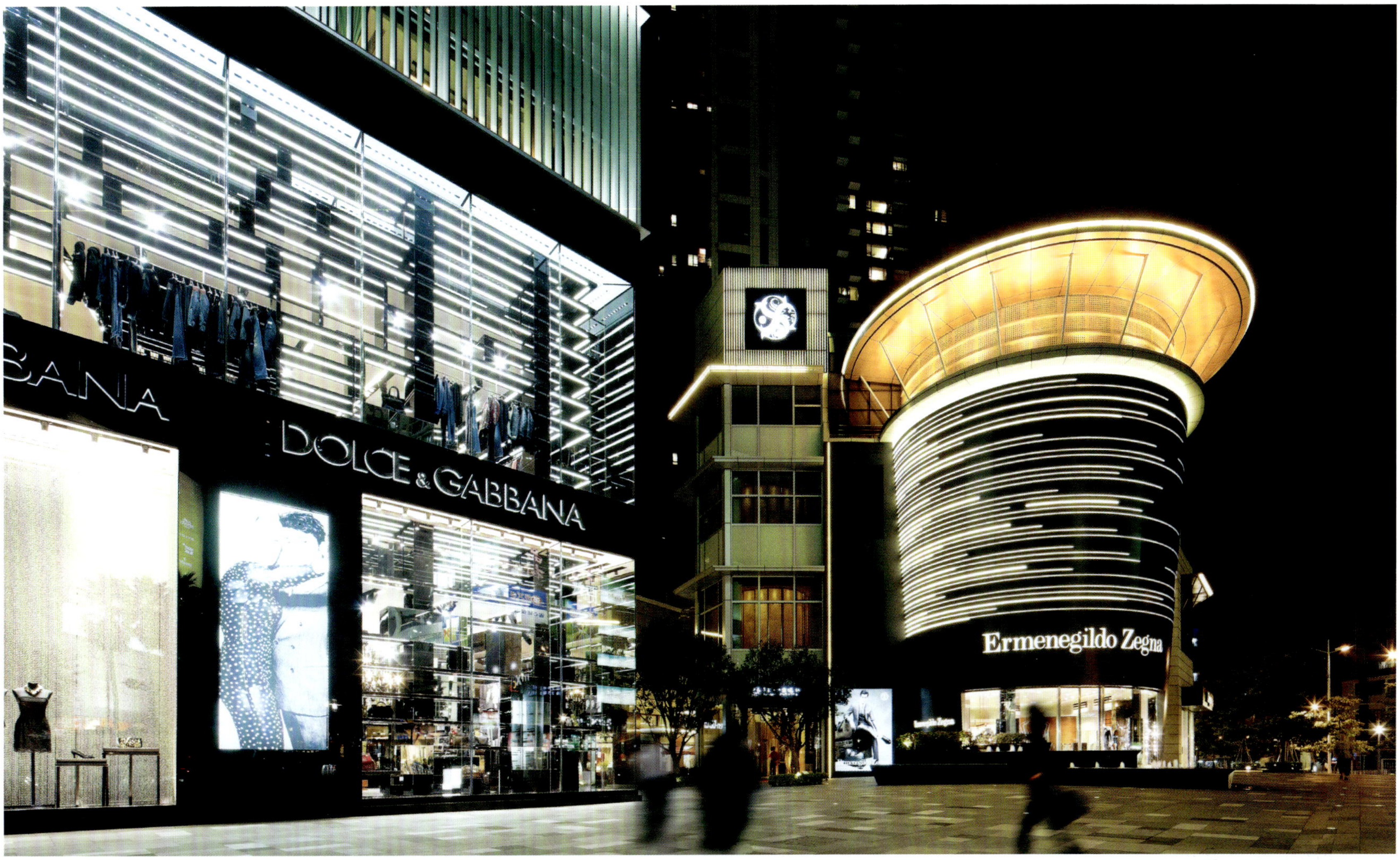
DOLCE & GABBANA
Ermenegildo Zegna

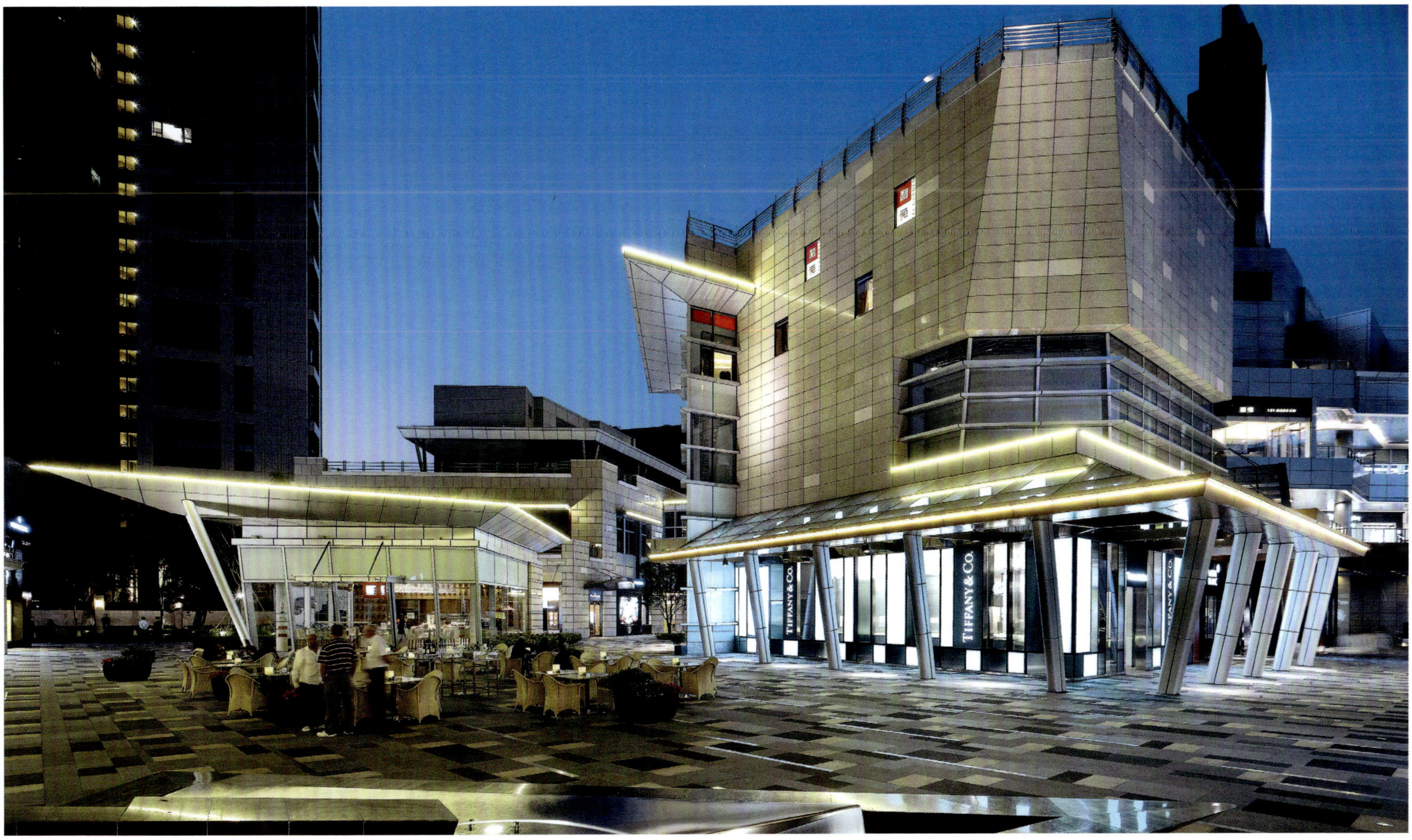
TIFFANY & Co.
TIFFANY & Co.

LOUIS VUITTON
Dior
BAO AN ROAD
SHENNAN ROAD
OFFICE TOWER
SHOPPING MALL
JINFENG
CITY
SHU CHENG ROAD
N-1 ROAD
JIN
SHAN
金山大厦

MONT
BLANC

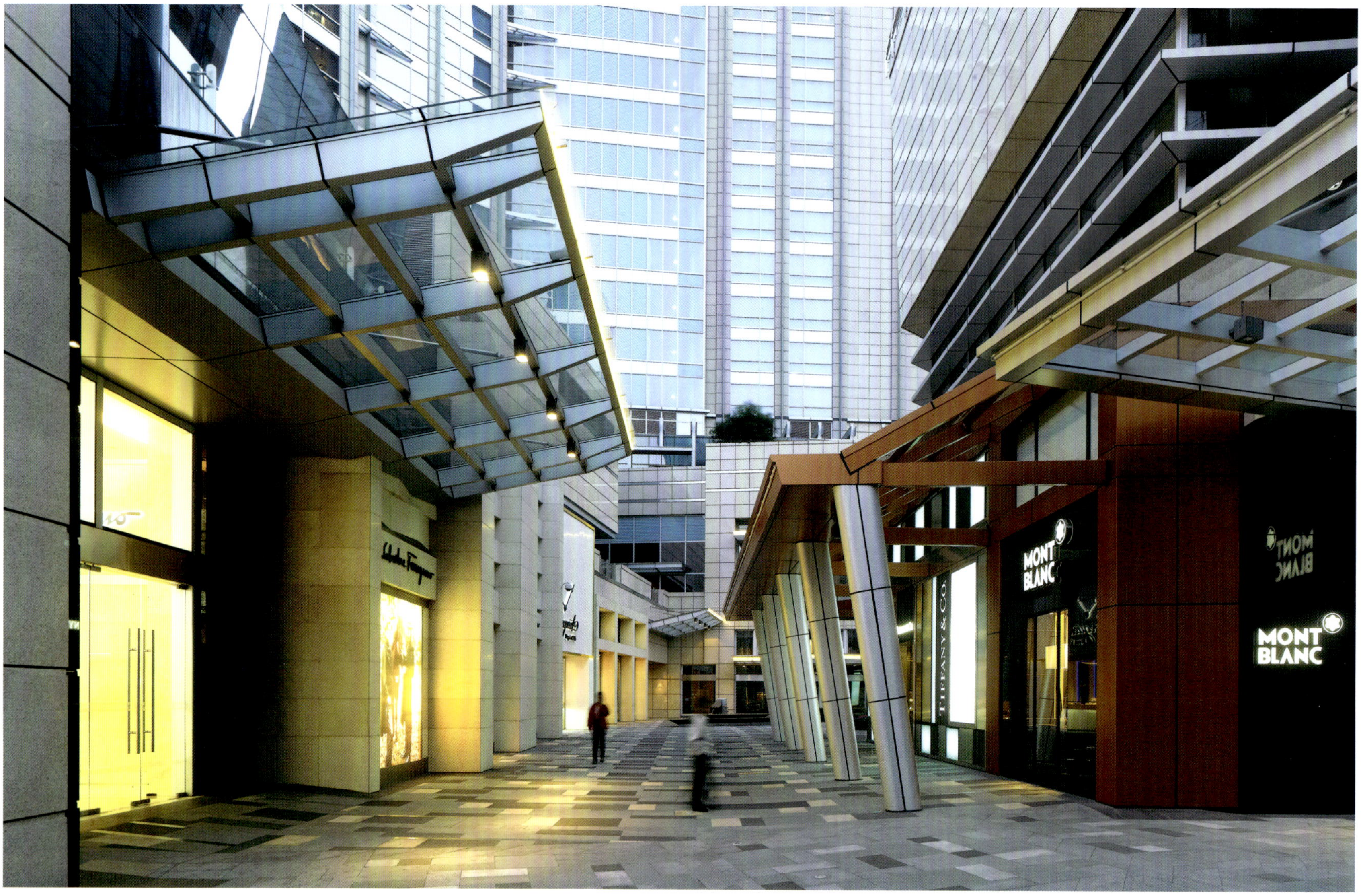
MONT BLANC
TIFFANY & CO.
MONT BLANC

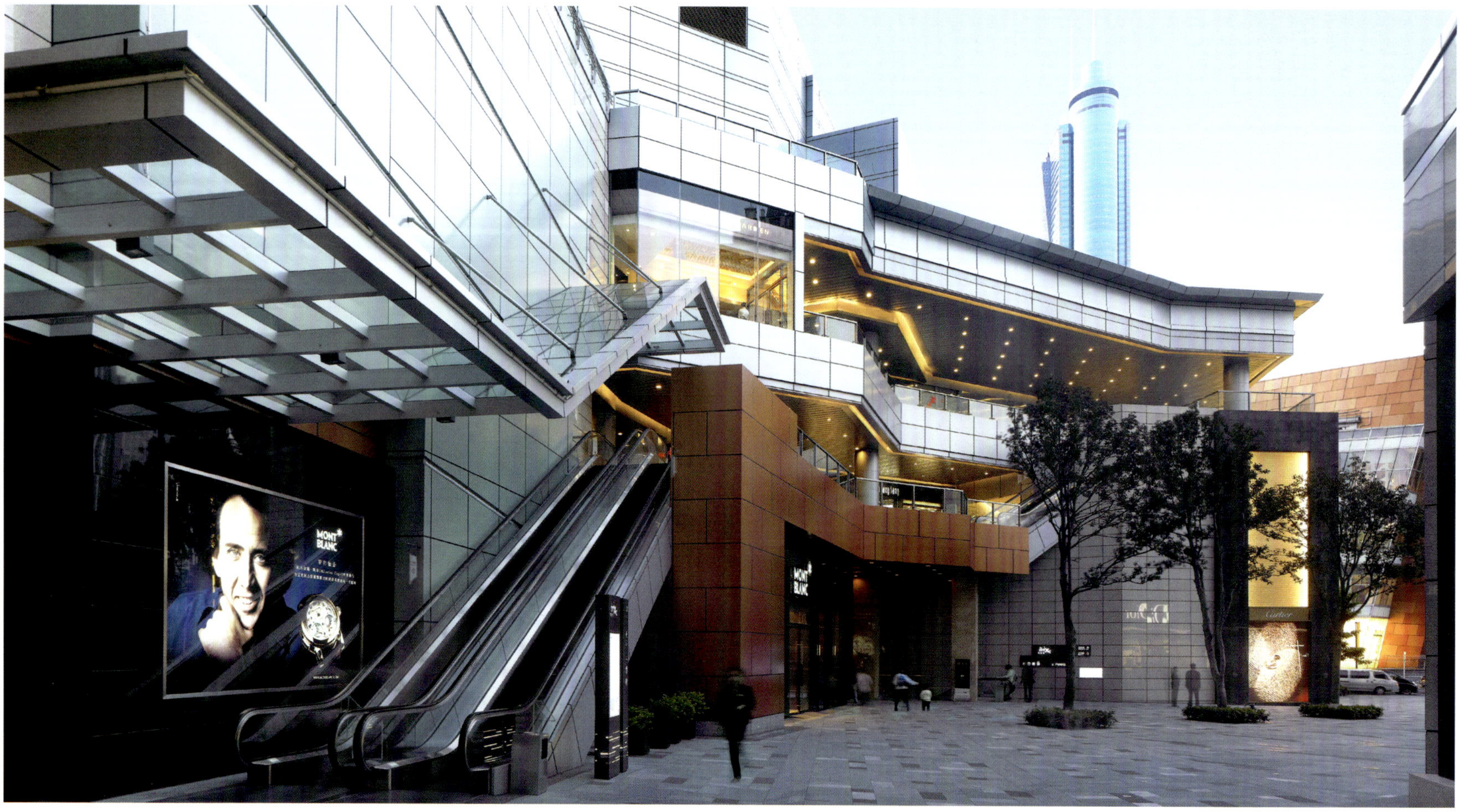
MONT BLANC

Cartier

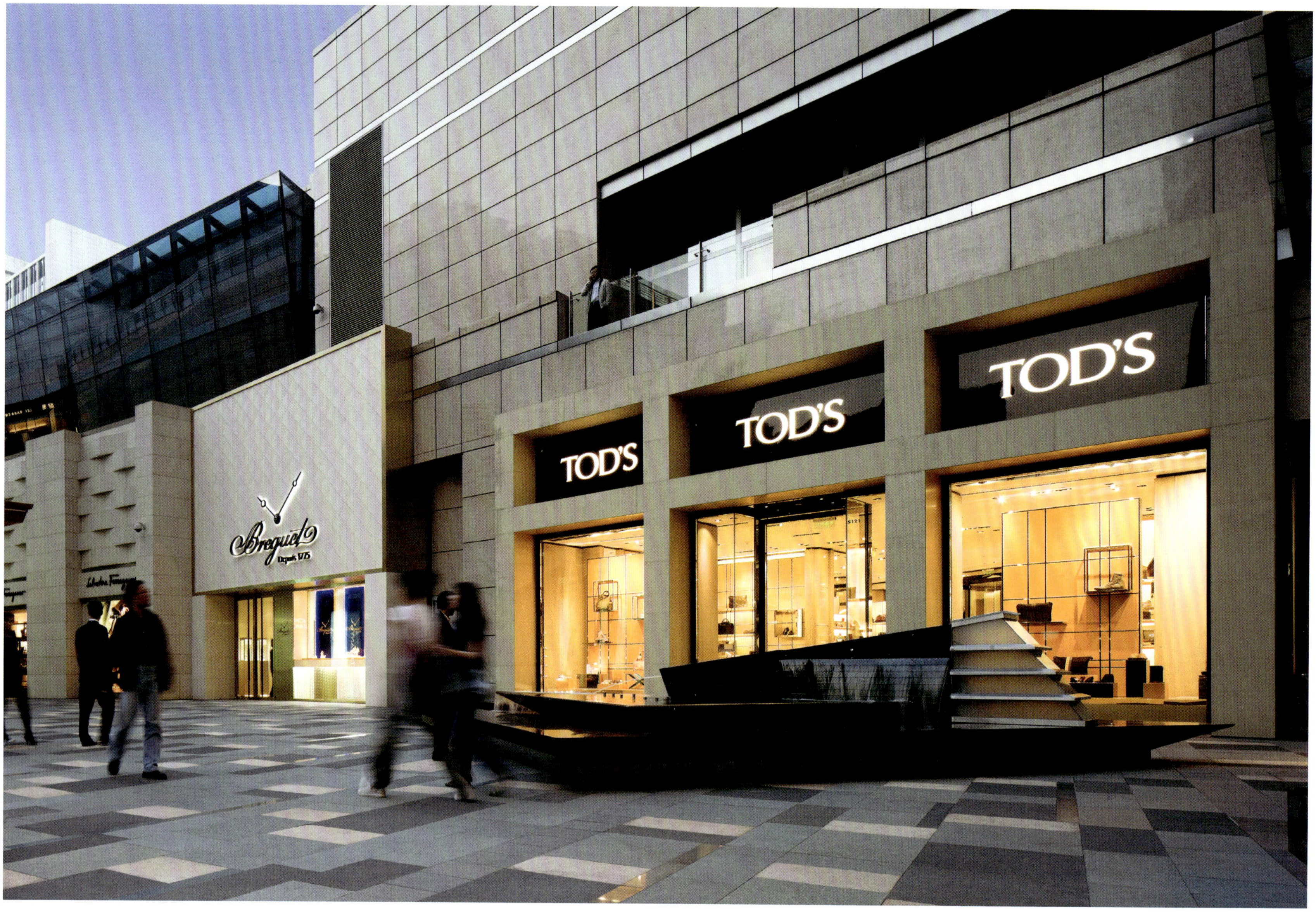

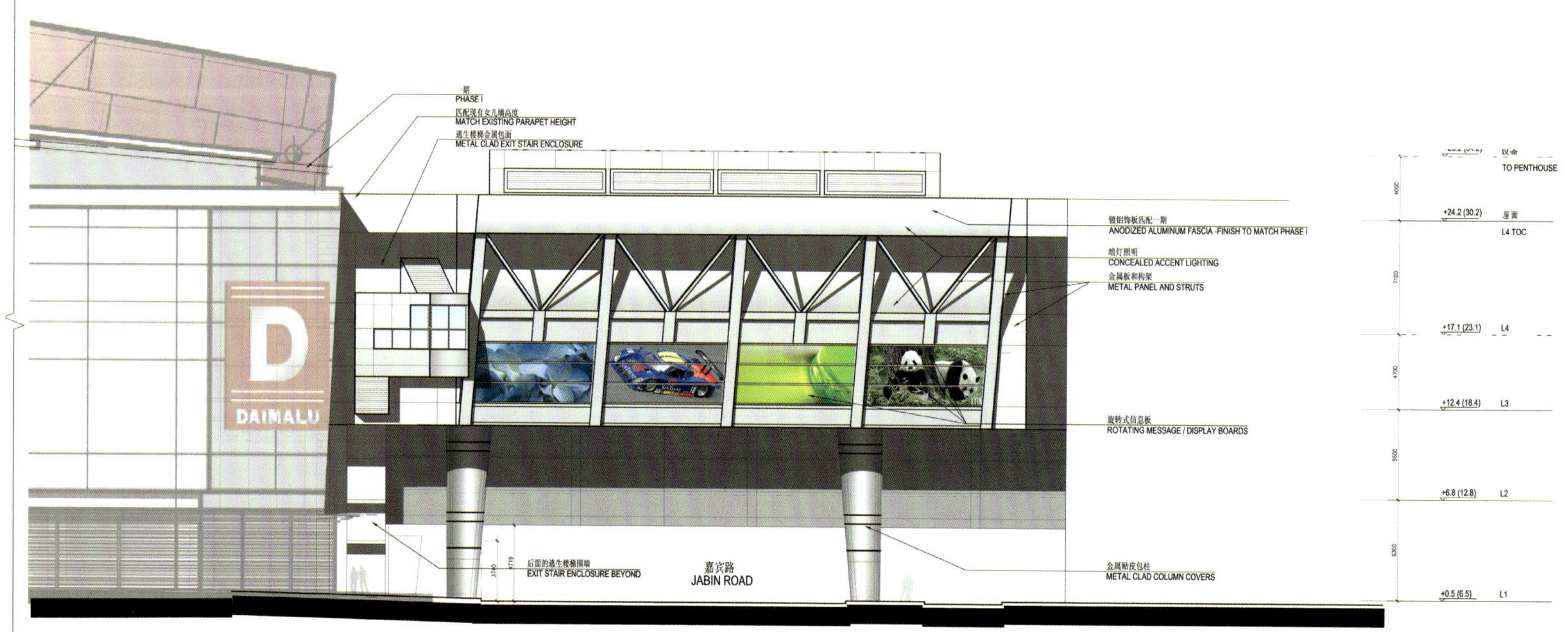
一期
PHASE I
匹配现有女儿墙高度
MATCH EXISTING PARAPET HEIGHT
逃生楼梯金属包面
METAL CLAD EXIT STAIR ENCLOSURE
镀铝饰板匹配一期
ANODIZED ALUMINUM FASCIA -FINISH TO MATCH PHASE I
暗灯照明
CONCEALED ACCENT LIGHTING
金属板和构架
METAL PANEL AND STRUTS
旋转式信息板
ROTATING MESSAGE / DISPLAY BOARDS
后面的逃生楼梯围墙
EXIT STAIR ENCLOSURE BEYOND
嘉宾路
JABIN ROAD
金属贴皮包柱
METAL CLAD COLUMN COVERS
DAIMALU
TO PENTHOUSE
+24.2 (30.2) 屋面
L4 TOC
+17.1 (23.1) L4
+12.4 (18.4) L3
+6.8 (12.8) L2
+0.5 (6.5) L1
4000
7100
4700
5600
5300
3740
4719

P9
P9
MTL-3
MTL-3
PAINTED METAL P13
METAL CANOPY MTL-1
METAL CORNICE MTL-3
PAINTED METAL FRAME P10
PAINTED METAL P13
STONE COPING G13
MTL-3
MTL-3
STONE TRANSITION LEDGE G3
STONE PANEL VENER G4-B
SEE 9G/31.45 FOR CONTINUED ELEVATION
PAINTED METAL P13
PAINTED METAL FRAME P10
PAINTED METAL CANOPY, BRACKETS & FRAME P7
PAINTED METAL CANOPY, & FRAME P10
PAINTED METAL COLUMN P10
GRANITE BASE G6
PAINTED METAL STORE FRONT P10
GRANITE BASE G6
+25.2 (31.2)
+22.2 (28.2)
+17.7 (23.7) L4
+12.7 (18.7) L3
+6.2 (12.2) L2
-0.2 (5.8) L1

TOP OF PENTHOUSE
L4 ROOF
L3
L2
L1

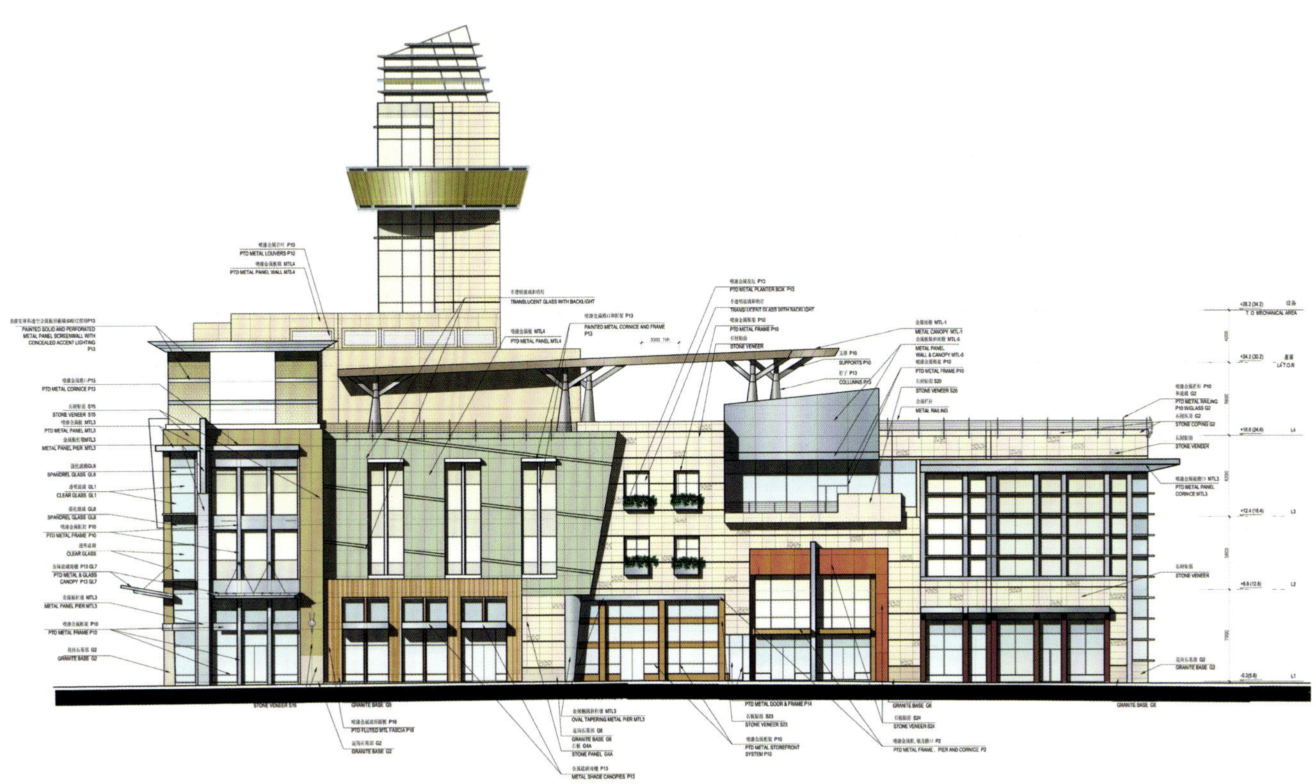

# Mirdif City Centre

## Mirdif城市中心

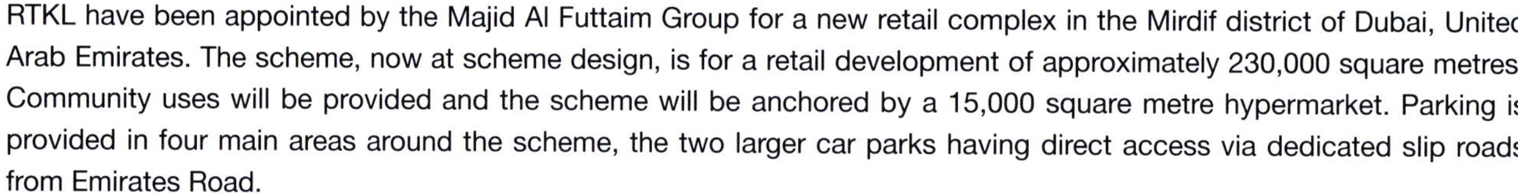

RTKL have been appointed by the Majid Al Futtaim Group for a new retail complex in the Mirdif district of Dubai, United Arab Emirates. The scheme, now at scheme design, is for a retail development of approximately 230,000 square metres. Community uses will be provided and the scheme will be anchored by a 15,000 square metre hypermarket. Parking is provided in four main areas around the scheme, the two larger car parks having direct access via dedicated slip roads from Emirates Road.

The project is located between Dubai International Airport and a largely residential area called Mirdif. The retail mall is on two levels and is planned on a rectangular layout with nodes at the intersections. These nodes connect, via welcome halls, to provide direct access from the car park decks. The most dramatic feature is the main central street which bisects the scheme. This public space is designed as an active urban zone and is configured as a series of interconnecting rooms that progress from the landscaped Mirdif "front door" through the heart of the scheme. These rooms are triple height spaces, eccentric to the central axis and express themselves as dramatic forms on the roofscape. The central plaza is the most prominent and is designed to accommodate large promotional events.

**Location**
Dubai, United Arab Emirates

**Area**
229,992.2513 sq.m.

**Company**
RTKL Associates Inc.

**Company**
RTKL Associates Inc.

**Photographer**
Mitch Duncan

The Majid Al Futtaim Group已委托RTKL在阿拉伯联合酋长国迪拜Mirdif区建立一个新的零售综合大楼。该计划，现在已经在计划设计中，是建立一个约23万m²的零售发展区。政府提供了公共用地，在这块土地上将规划歀一个15,000m²的大型超市。该计划周围的4个主要区域用于建设停车场，从阿联酋路经过专用支路可以直接进入2个较大的停车场。

该项目位于迪拜国际机场和一个名叫Mirdif的主要住宅区之间。零售商场共2层，计划采用在交叉口布有节点的矩形布局。这些节点通过迎宾大厅相互连接，从停车场甲板开始提供了直接进入商场的入口。最显着的特点是将该方案一分为二的主要的中央大街。这个公共空间设计成为一个活跃的城市区域，并配置一系列相连的房间，这些房间从风景如画的Mirdif“前门”穿过该方案的核心。这些房间空间拥有三重高度，偏离中心轴和并且将其表达成屋顶景观上的戏剧形式。中环广场是最突出的设计，以用作大型宣传活动。

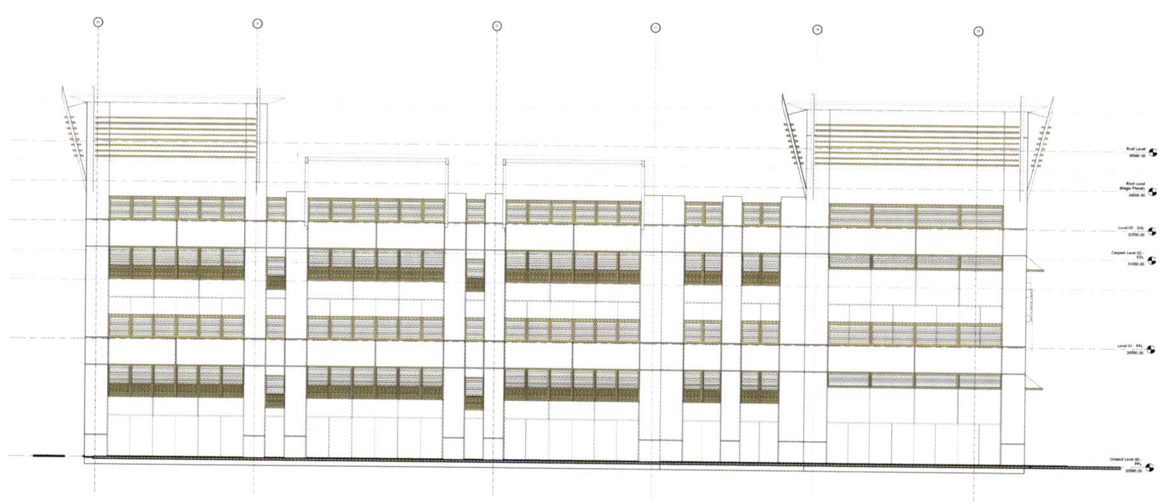

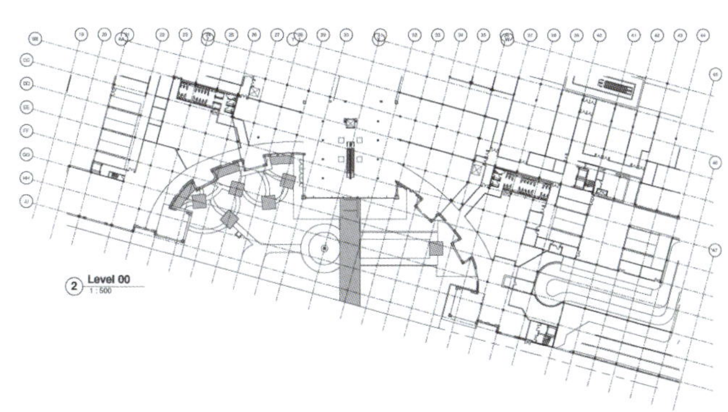

2 Level 00
1:500

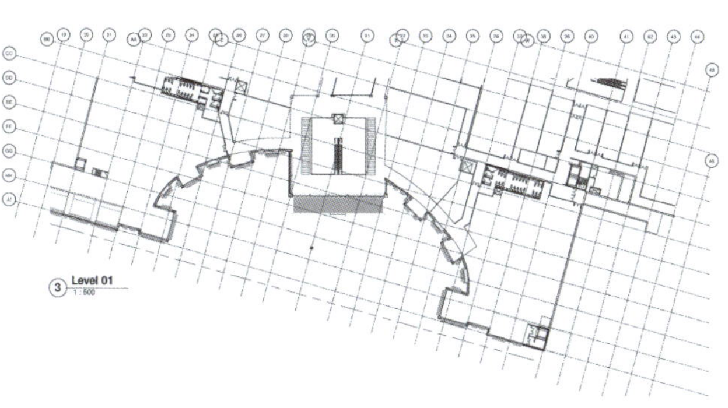

3 Level 01
1:500

SEE BY CHLOE
كرت جايجر
SHADES

8 Part Elevation
1 : 100

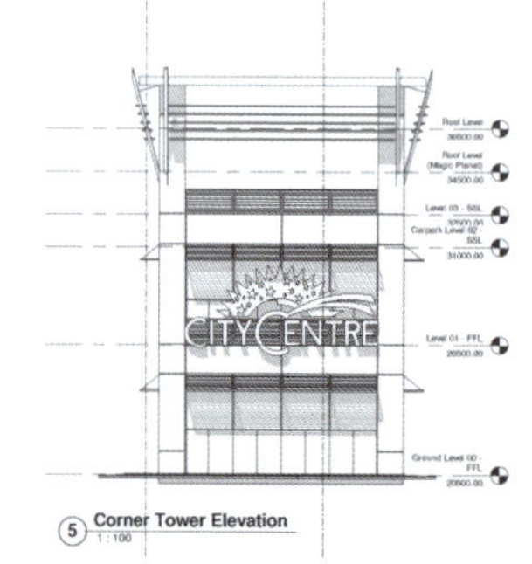


5 Corner Tower Elevation
1 : 100

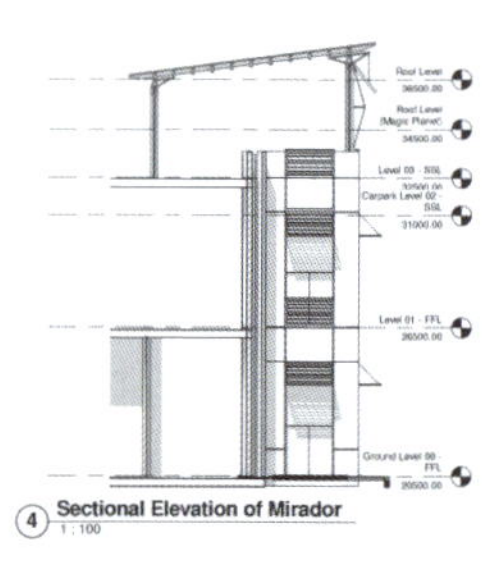

4 Sectional Elevation of Mirador
1 : 100

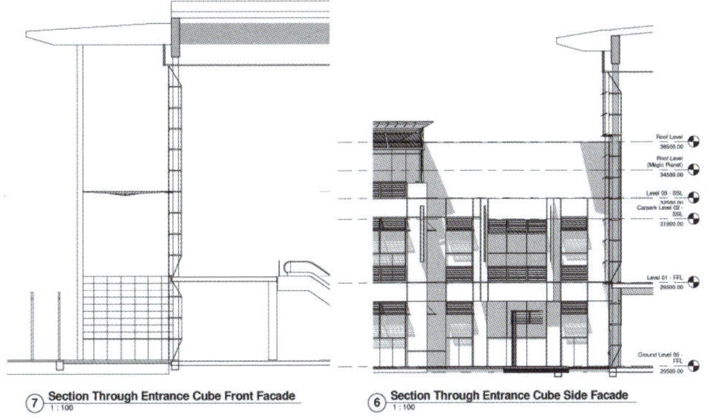

7 Section Through Entrance Cube Front Facade 1:100

6 Section Through Entrance Cube Side Facade 1:100

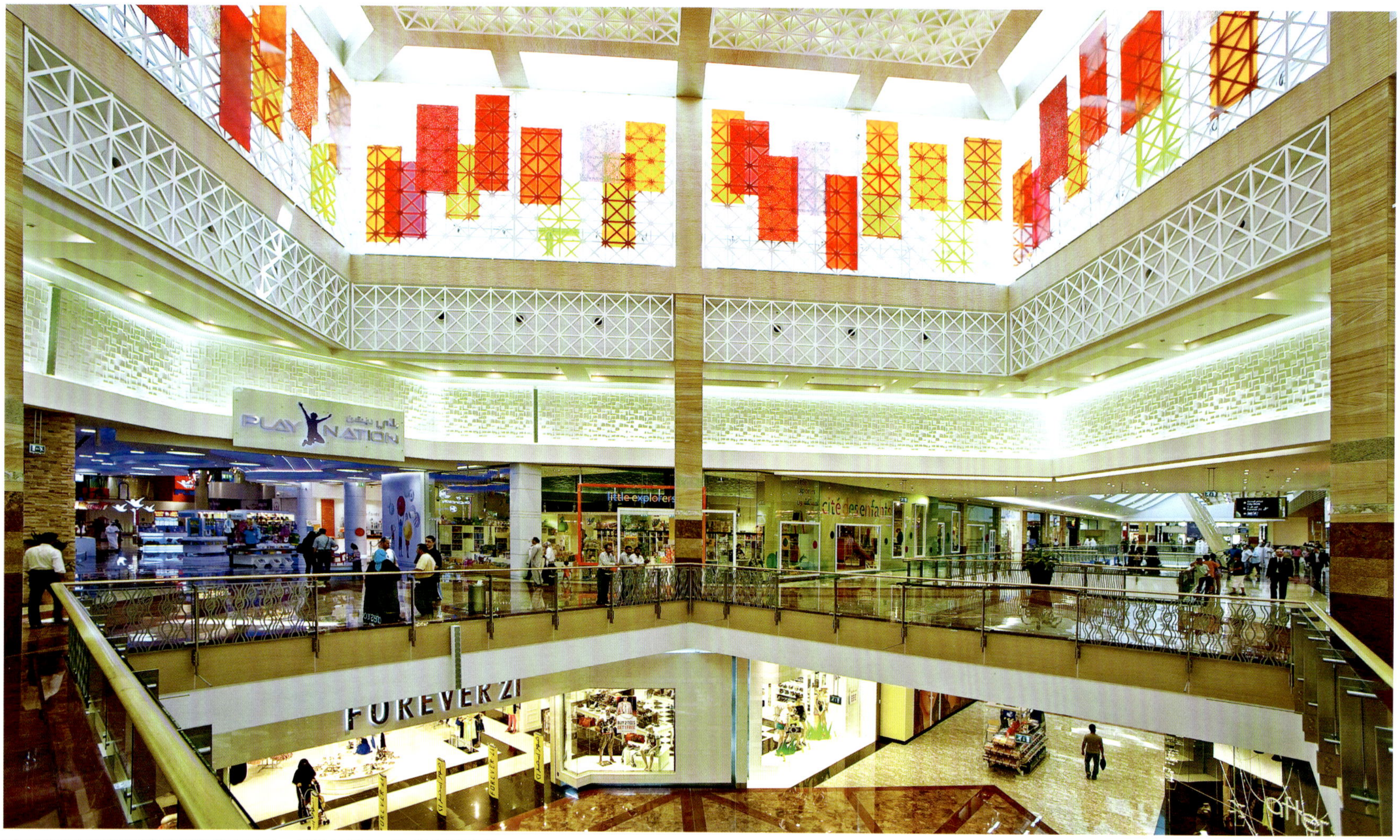
PLAY NATION
FOREVER 21

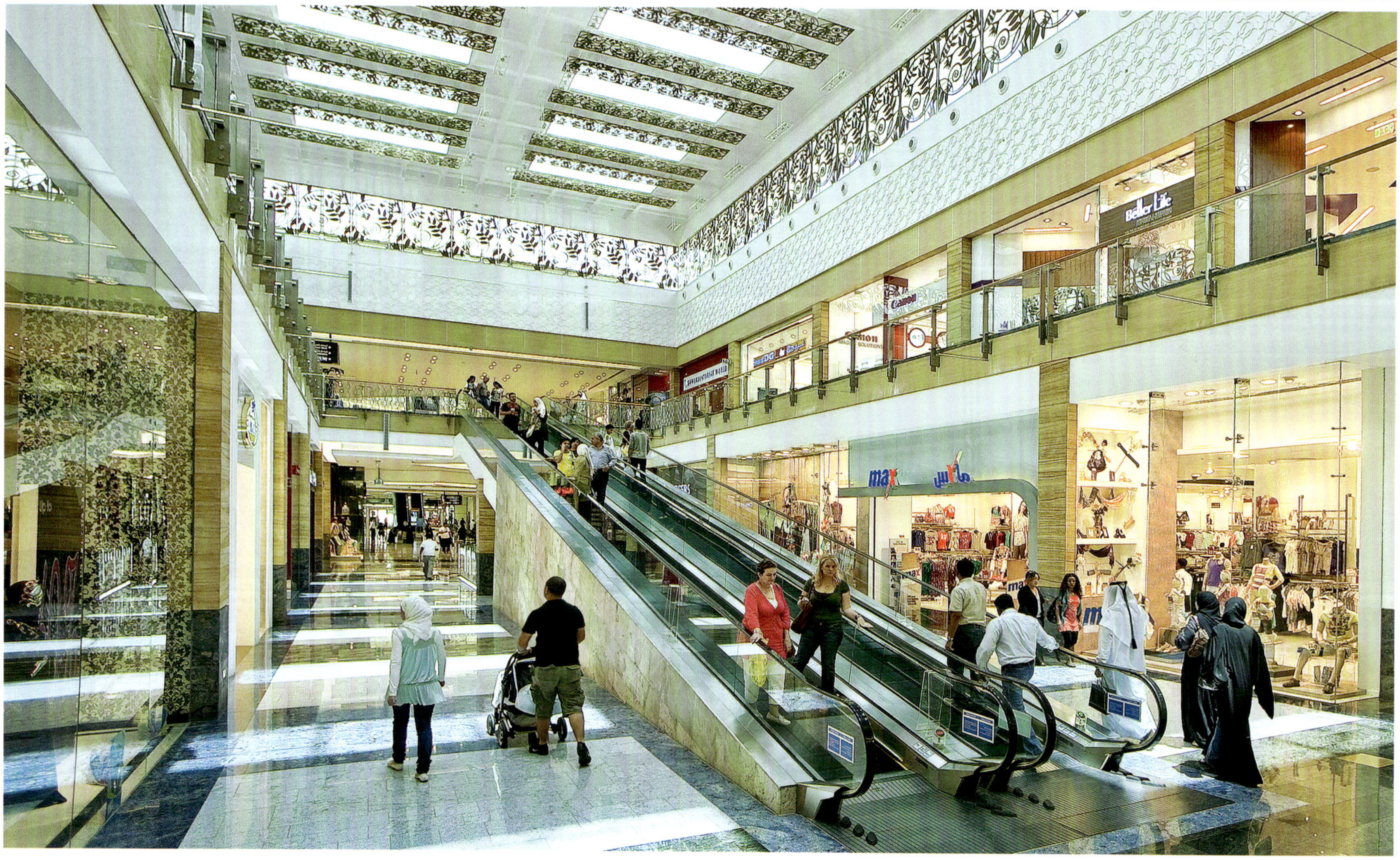
max

ecco
SALE
KIDS ONLY
HSBC

# John Lewis Department Store and Cineplex

# 约翰路易斯百货商场和电影院

Department stores are conventionally designed as blank enclosures to allow retailers the flexibility to rearrange their interior layouts. However, the physical experience of shops is an increasingly important consideration to compliment the convenience of online shopping. The concept for the John Lewis store is a net curtain, providing privacy to the interior without blocking natural light. The design of the store provides the retail flexibility required without removing the urban experience from shopping. The store cladding is designed as a double glazed façade with a pattern introduced, making it like a net curtain. This allows for a controlled transparency between the store interiors and the city, allowing views of the exterior and natural light to penetrate the retail floors whilst also future - proofing the store towards changes in layout. Thus, the store is able to reconfigure its interiors without compromising on its exterior appearance.

The pattern itself is formed of four panels of varying density which allow for a variable degree of transparency. These meet seamlessly across the perimeter, producing a textile - like cladding. Frit in mirror onto two layers of glass curtain wall, the mirrored pattern reflects its surroundings and in doing so becomes further integrated into its context, densifying and changing as the sun moves around the building. Viewed frontally from the retail floors, the double façade aligns to allow views out, whilst an oblique view from street level displaces the two patterns and creates a moiré effect, reducing visibility and increasing visual complexity, thereby maximizing the privacy performance.

**Location**
Leicester, UK

**Area**
34,000 sq.m.

**Main materials**
Concrete, glass, wood, etc.

**Company**
Foreign Office Architects

**Designers**
Farshid Moussavi, Alejandro Zaera-Polo

**Photographers**
Satoru Mishima, Peter Jeffree, Helene Binet, Lube Saveski

传统的百货商场被设计为空白围墙，可允许零售商灵活地重新规划他们的室内布局。然而，为了迎合网上购物的便利性，店铺的实际购物体验成为一个日益重要的考虑因素。约翰路易斯百货的理念为一面网状窗帘，在不阻挡自然光的同时也为室内提供隐私性。

百货的设计提供必要的零售灵活性，同时不剔除购物中的城市体验。百货的外墙被设计为带有花纹的双面玻璃门面，从而使它像一面网状窗帘。这使得商店内部与城市之间存在一种受控的透明度，使得外部风景和自然光渗透零售楼面，同时面对变化时百货的布局可以适应未来发展。因此，在不改变其外观的同时，百货可以重新配置室内装饰。

花纹本身由4块不同密度的面板组成，形成不同程度的透明度。穿过外围之后，它们无间隙地连接在一起，从而形成一面纺织品般的外墙。在两层的玻璃幕墙上熔入镜子，这种镜面图案反射了其周围环境，如此一来，进一步地融合成它的脉络，当太阳围绕着建筑转时，脉络更加紧密而且千变万化。从零售楼层的正面看，在外面可透过双层门面看到里面，同时街道层上倾斜的风景取代了两种图案，形成一种云纹效果，降低了可见度却增加了视觉复杂性，从而使隐私功能最大化。

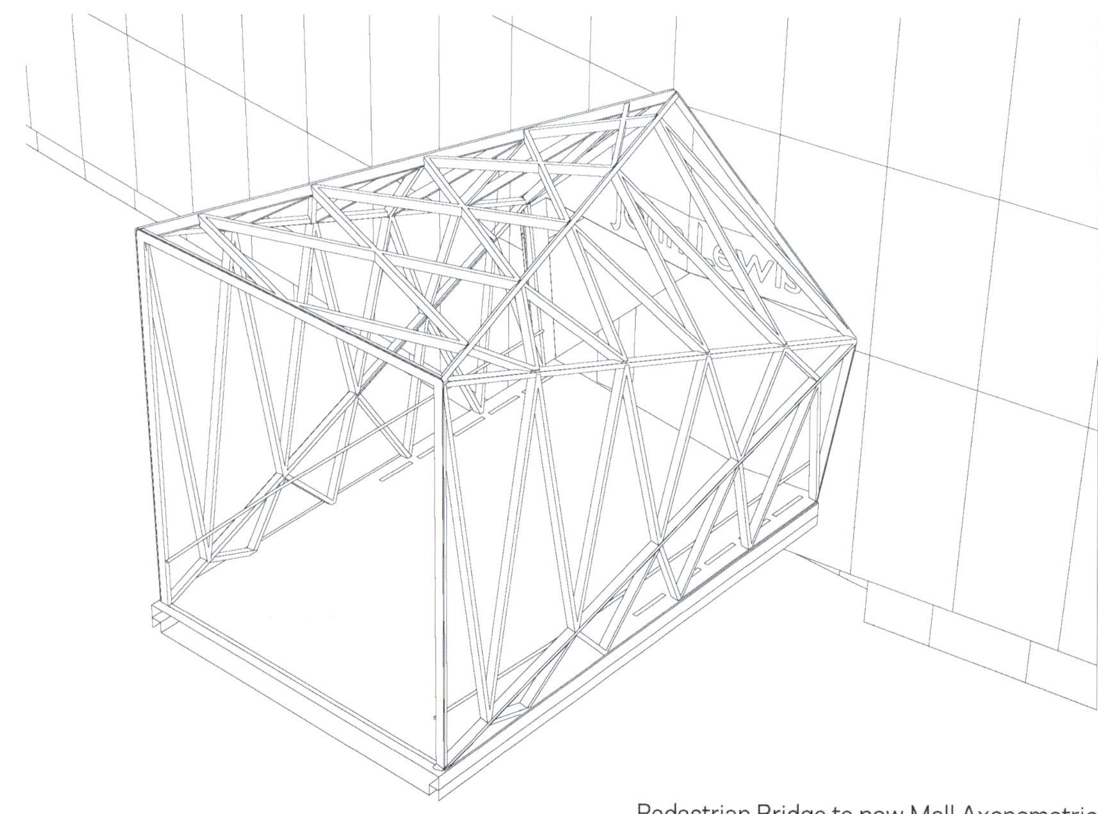

Pedestrian Bridge to new Mall Axonometric

# Sections

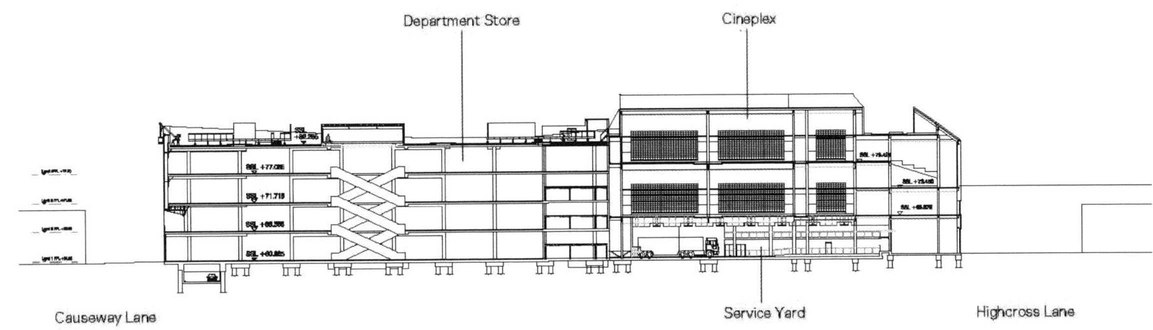


Longitudinal Section through Department Store and Cineplex

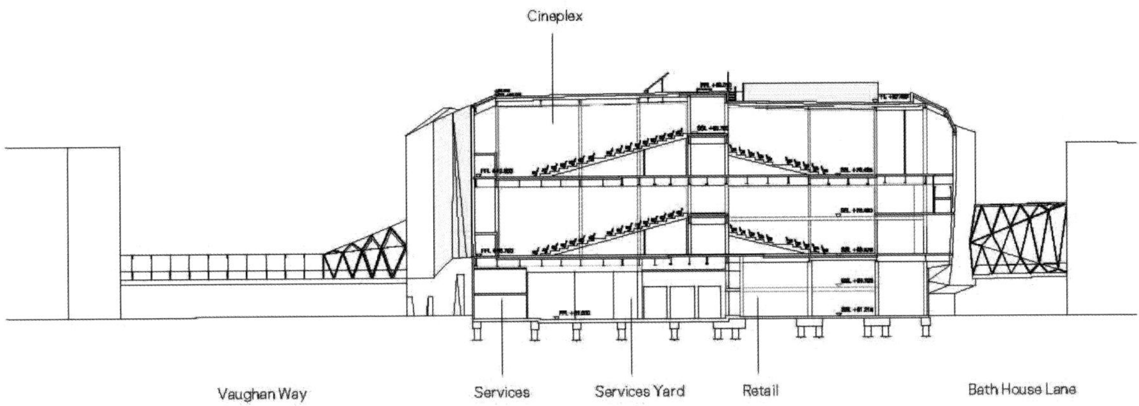


Tranversal Section through Cineplex

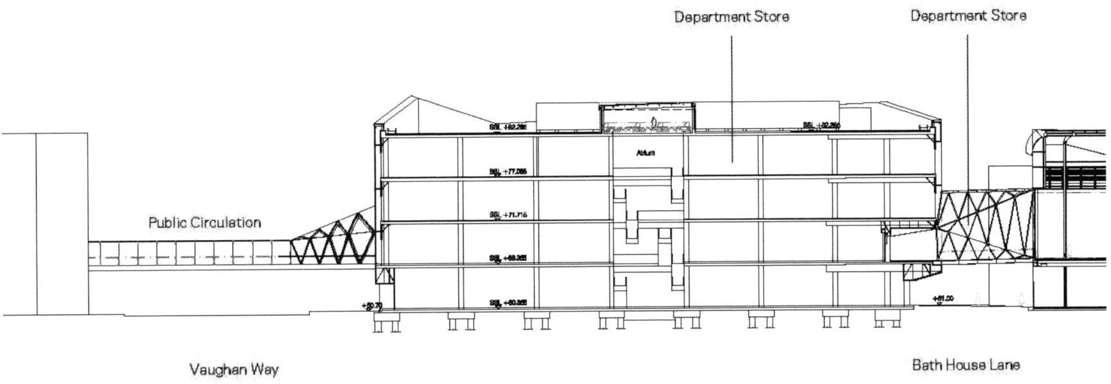


Tranversal Section through John Lewis

Sectional Axonometric through Department Store exterior envelope

Section through Vaughan Way pedestrian bridge

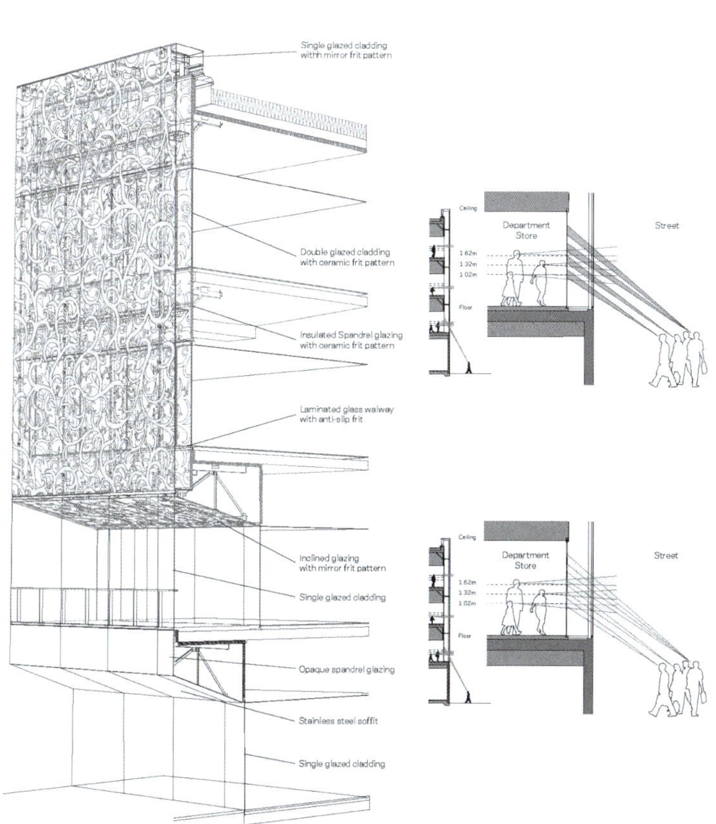


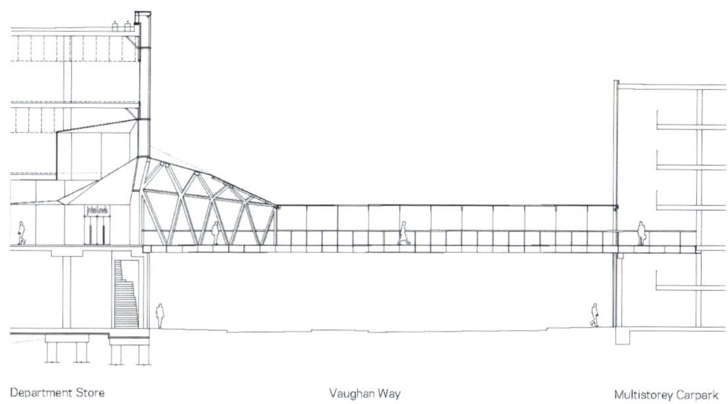


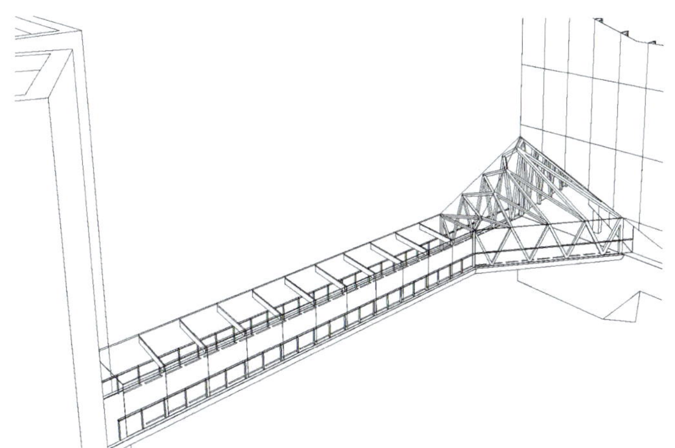

# Atakoy Plus Shopping Centre

## Atakoy Plus购物中心

Atakoy Plus [A+] Shopping Center is located in Atakoy, a district of Istanbul, Turkey. Atakoy Plus Shopping Center provides an alternate solution to the standard shopping experience and arrives at a contemporary solution to a rapidly evolving building type in Istanbul.

The new shopping center is located in the heart of a growing residential community. Atakoy Plus Shopping Center announces itself through the complexity of its building facade, providing a glimpse of the retail stores within, while also serving as a constantly evolving billboard to address the changing environment within.

GAD's architectural strategy plays out in the building's facade. The facade of Atakoy Plus Shopping Center performs as a layering of alternating surfaces, including a wire mesh skin wrapping the rigid steel & glass inner structure. The facade performs differently, depending on its north, south, east and west orientation. The building's facade also responds to the urban fabric surrounding the building, which varies from residential on one side, to a more urban commercial district on the other sides.

Thus, the program for Atakoy Plus Shopping Center is arranged to actively engage both the customers and merchants within, while also announcing itself to the community on the exterior as a unique and modern shopping experience.

**Location**
Atakoy, Istanbul, Turkey

**Area**
97,331 sq.m.

**Main Materials**
Steel-ferro concrete, glass, etc.

**Company**
Global Architectural Development

**Photographer**
Ozlem Avcioglu

Atakoy Plus［A+］购物中心位于土耳其伊斯坦布尔的一个区：阿塔科伊。Atakoy Plus购物中心为普通的购物体验提供了一个可选择的解决方案，并且为伊斯坦布尔迅速发展的建筑类型提供了一个当代的解决办法。

新的购物中心位于发展中的住宅区中心。Atakoy Plus购物中心通过其复杂的建筑外立面展现自己，不仅使人可以瞥见里面的零售店，同时还可作为一个不断发展变化的广告牌，以适应不断变化的内环境。

GAD的建筑策略完全用在建筑的立面上。Atakoy Plus购物中心的立面是交替表面的分层，包括一个包裹着硬钢与玻璃内部结构的铁丝网状皮。在它的北边、南边、东边和西边，建筑立面都表现不同。建筑的立面也响应了围绕着该建筑的城市结构，一边是住宅区，另一边是一个更都市化的商业区。

因此，Atakoy Plus购物中心这个项目被用来主动吸引消费者和里面的商人，同时也在外观上向社区宣称它是一个独特的现代购物体验。

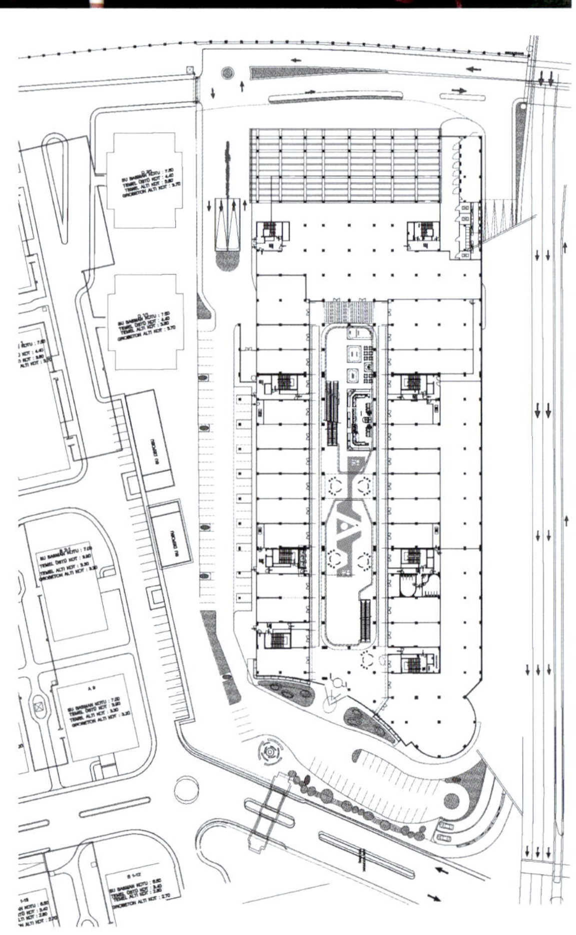

SHAFT
MECHANICAL ROOM
CINEMA
SERVICE CORRIDOR
SHOP
GALLERY SPACE
PROMOTION CORRIDOR
FOYER
BRIDGE
CINEMA ENTRY
ELECT ROOM
WC MAN
WC WOMAN
OFFICE
HALF OPEN PLACE

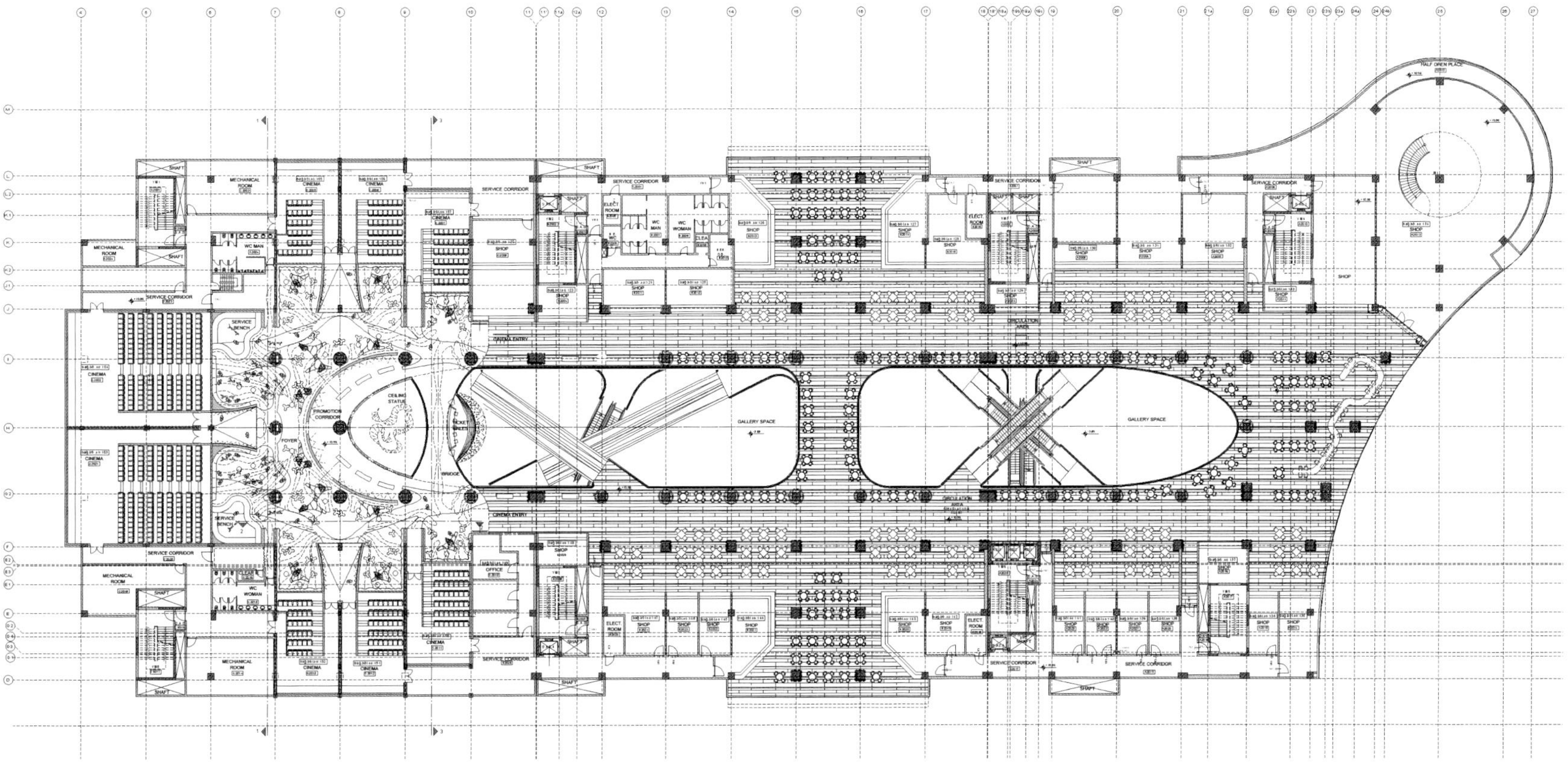

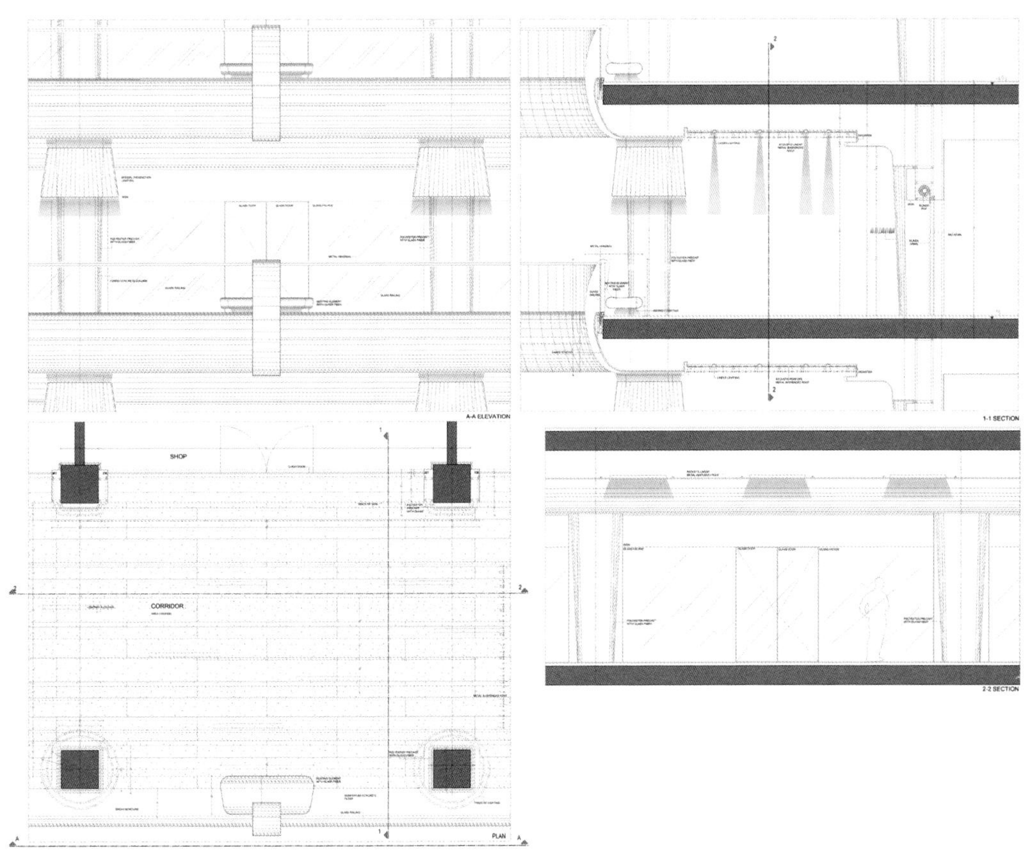

A-A ELEVATION
1-1 SECTION
SHOP
CORRIDOR
PLAN
2-2 SECTION

MAC
YARGICI
TERSANE
indirim
MM MiGRO
INDIRIMDE
70
Backhaus
Garden Salad

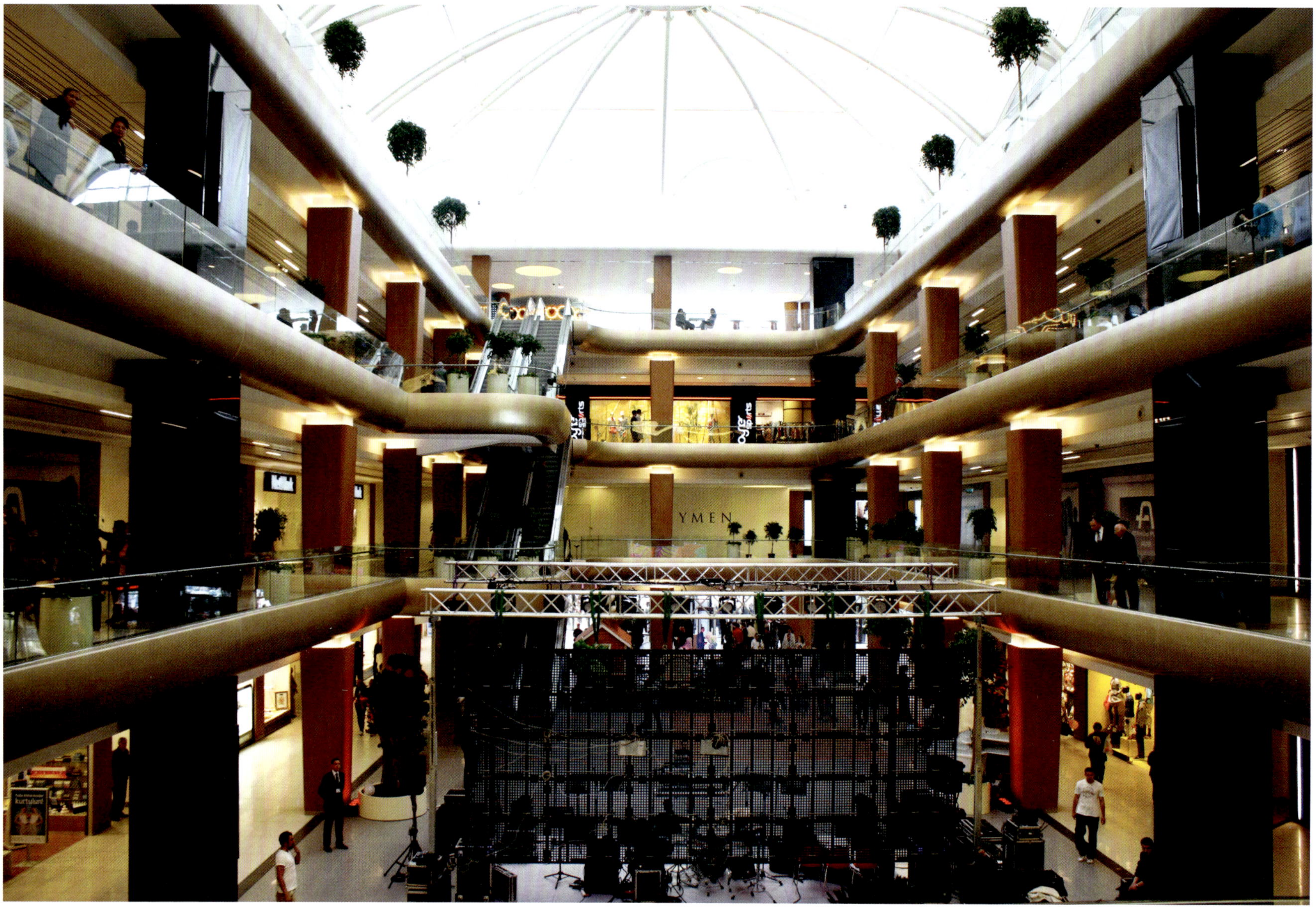
YMEN

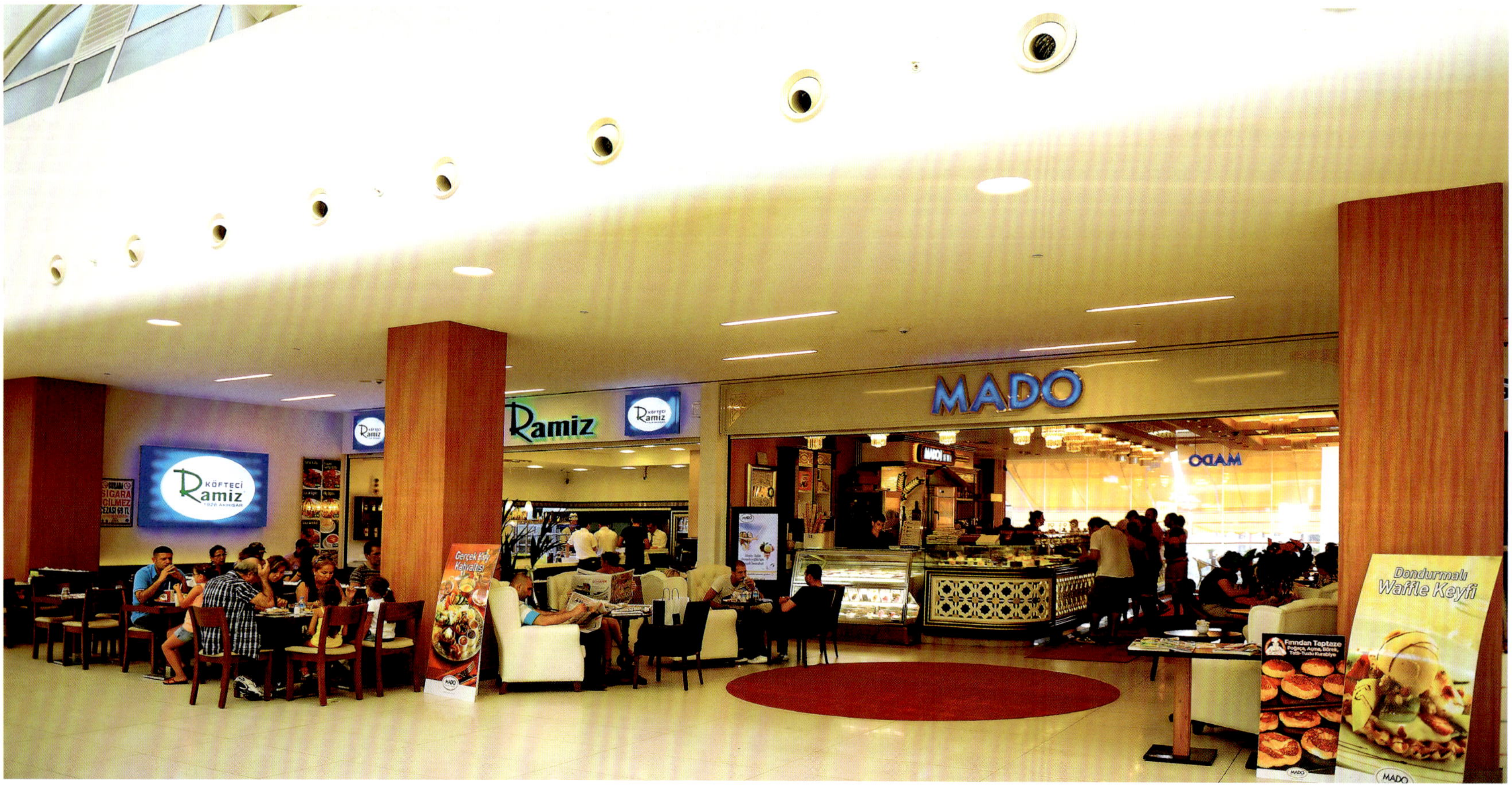
MADO
Ramiz
Dondurmalı
Waffle Keyfi

# Q19

## Q19购物中心

Q19 is the name of an attractive shopping center the charm of which lies in the fusion of old and new: the historically protected Samum paper factory at the northern end of the well - known Karl - Marx - Hof in Viennas 19th district is linked to a new construction. Advertising, facade and an integrated forecourt unify into an urban public space. Clients make their way through different exciting rooms, in which the interplay of daily and artificial light creates a kind of city center atmosphere.

Q19 是一个极具吸引力的购物中心，其魅力在于该项目的设计融合了新旧元素：这个历史保护的Samum造纸厂，南邻Vienas 19区的知名的的卡尔·马克思·霍夫，又与一栋新建筑相连。广告、门面和综合前院共同铸造了一个城市公共空间。客户穿梭在各种不同的令人兴奋的商店里，日光与人造光交相辉映，共同营造出一种城市中心的氛围。

**Location**
Vienna

**Company**
Peterlorenzateliers ZT GmbH

**Designer**
Peterlorenzateliers ZT GmbH

Q
19

OSTANSICHT

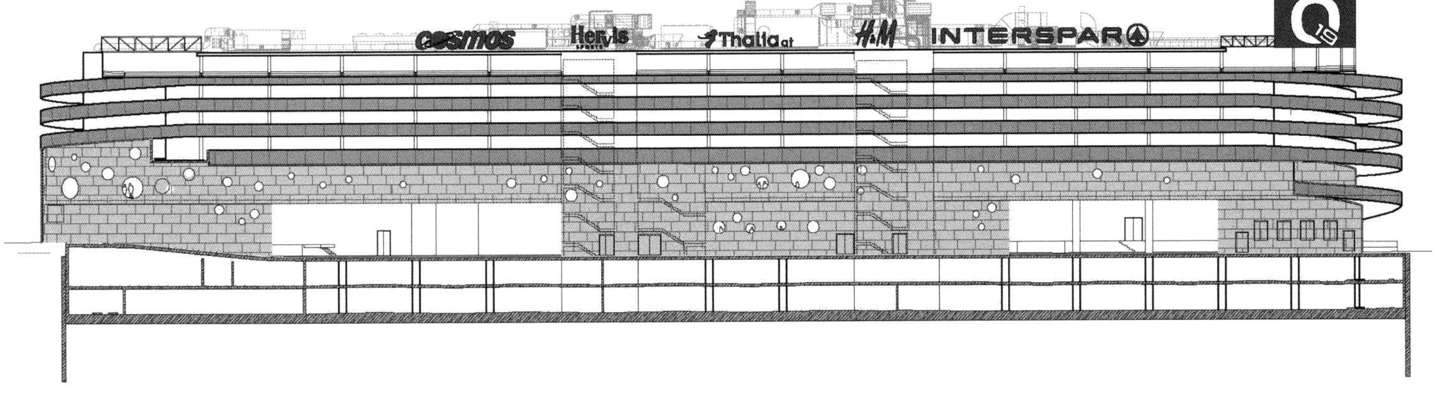


SÜDANSICHT

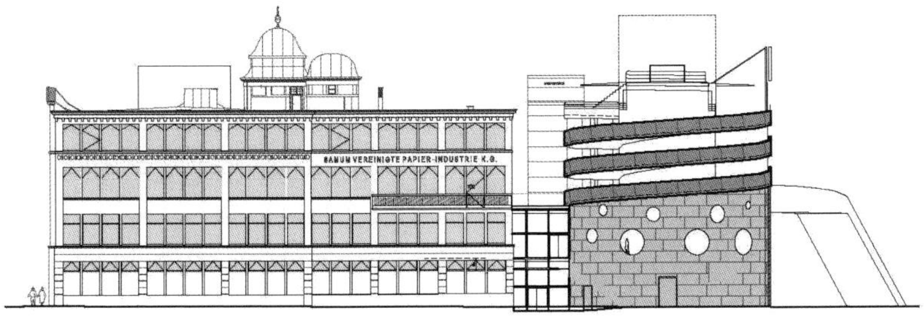

NORDANSICHT

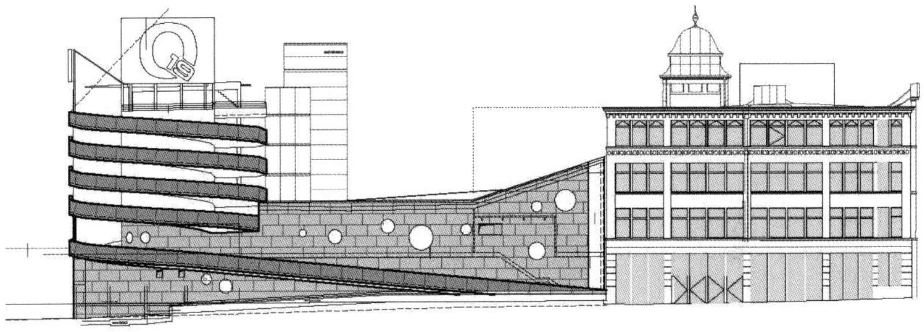

SCHNITT 5-5

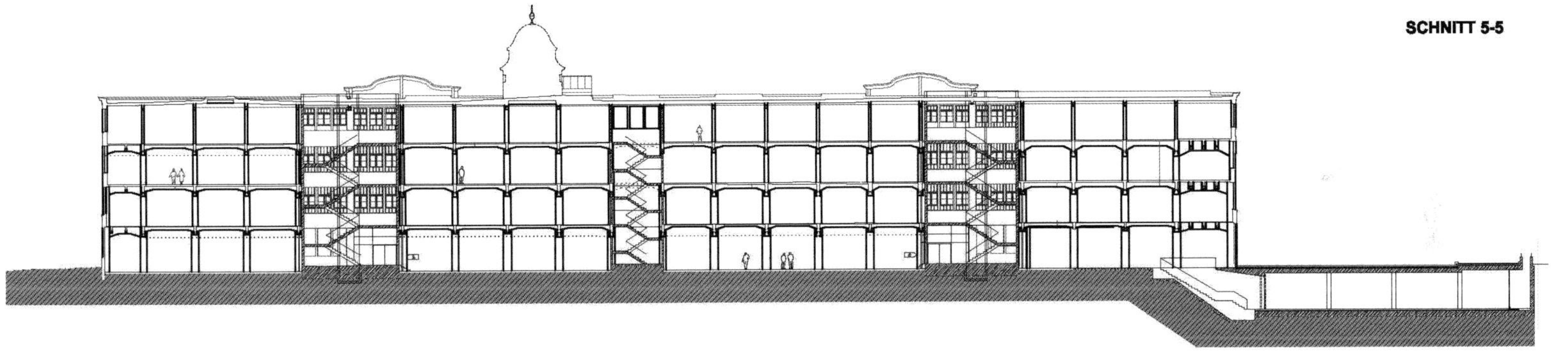

OSTANSICHT ALTBAU

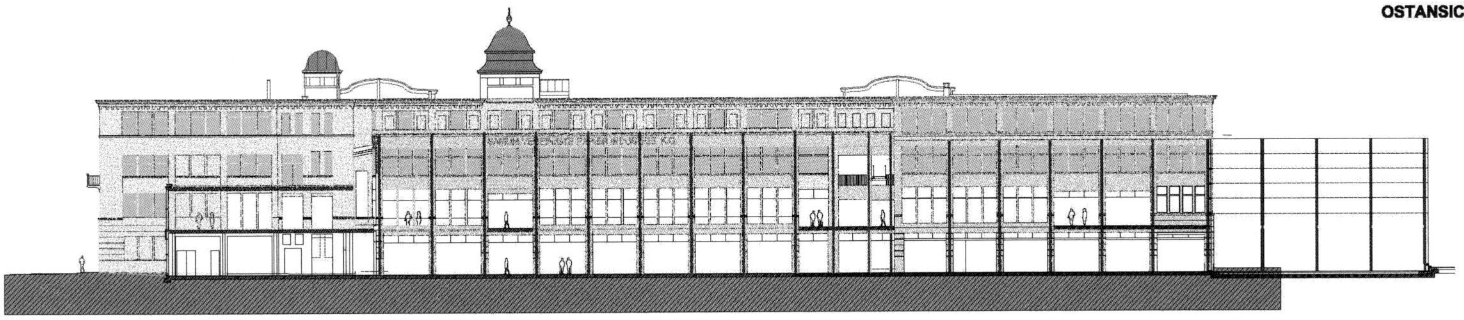

WESTANSICHT

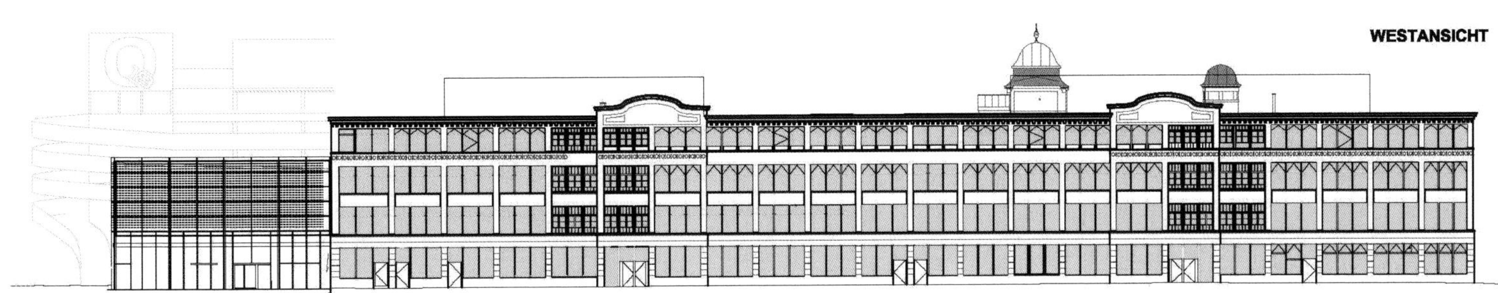

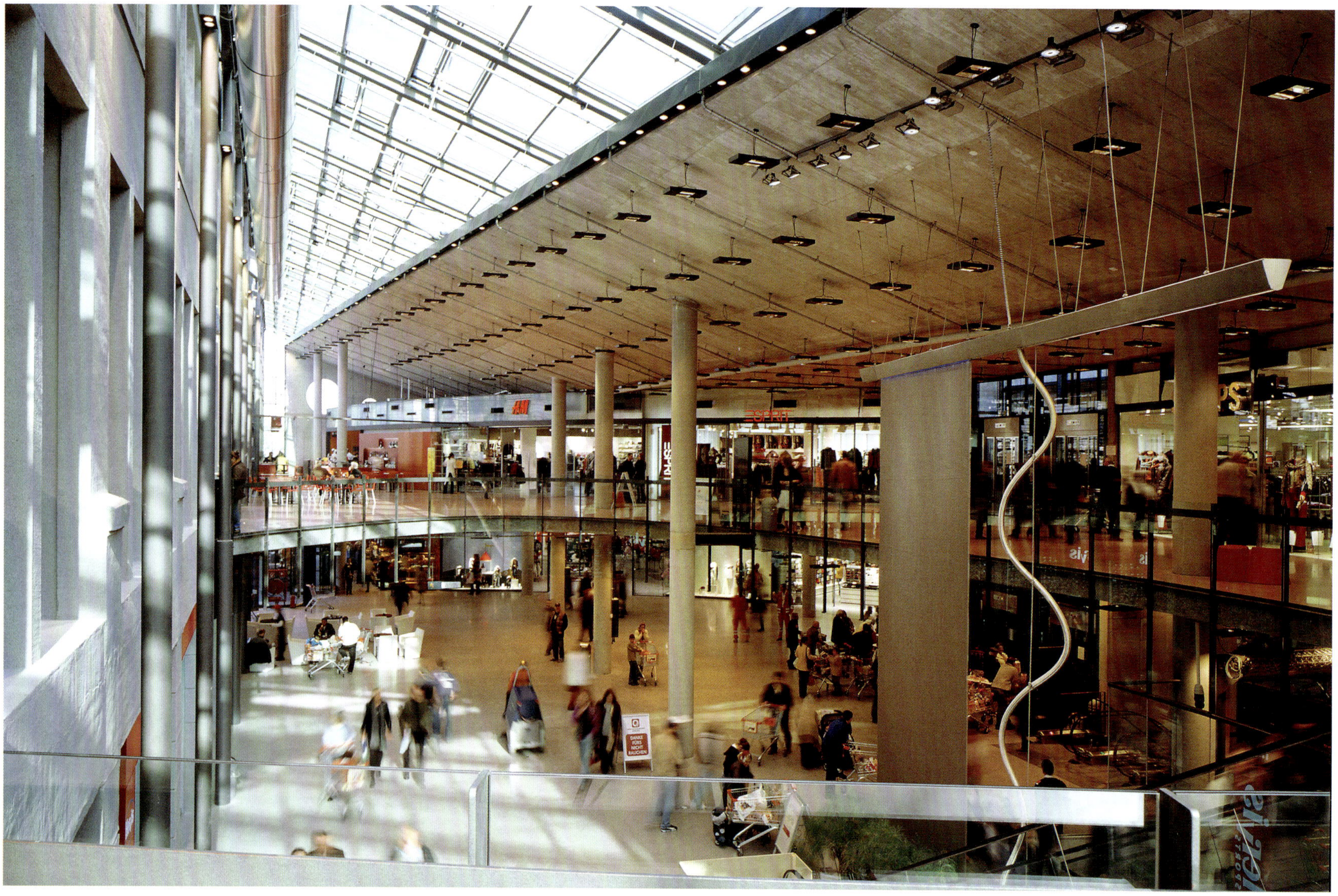

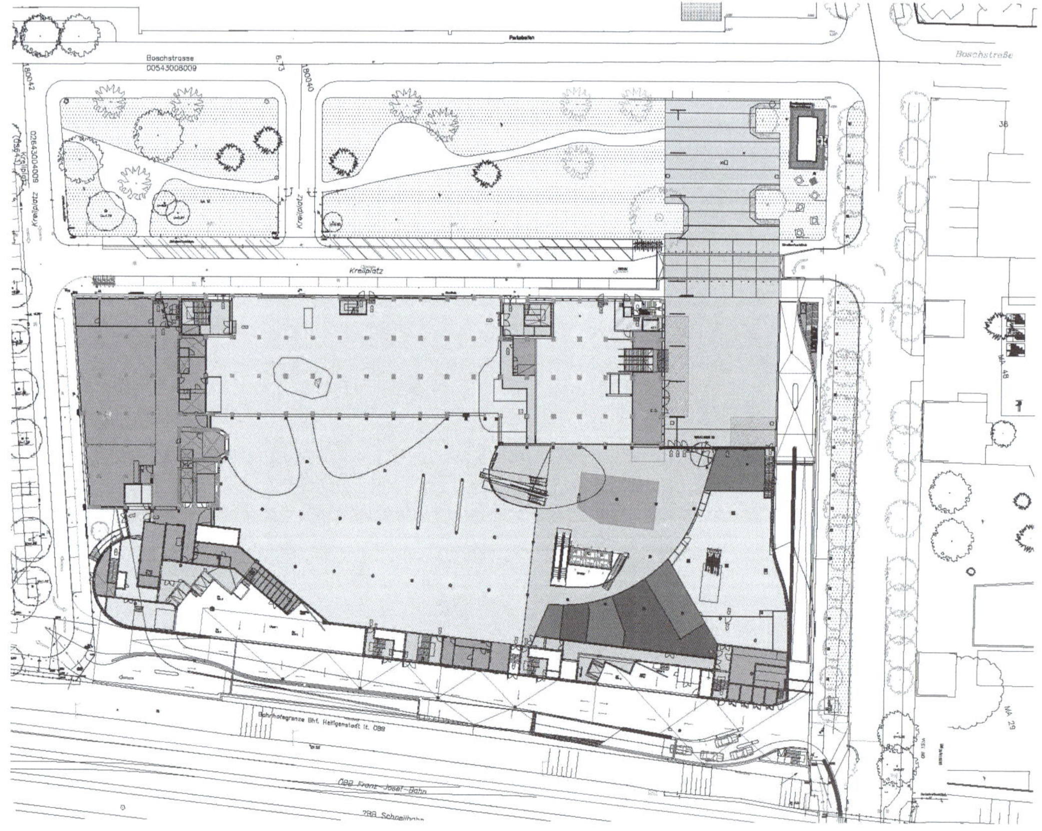

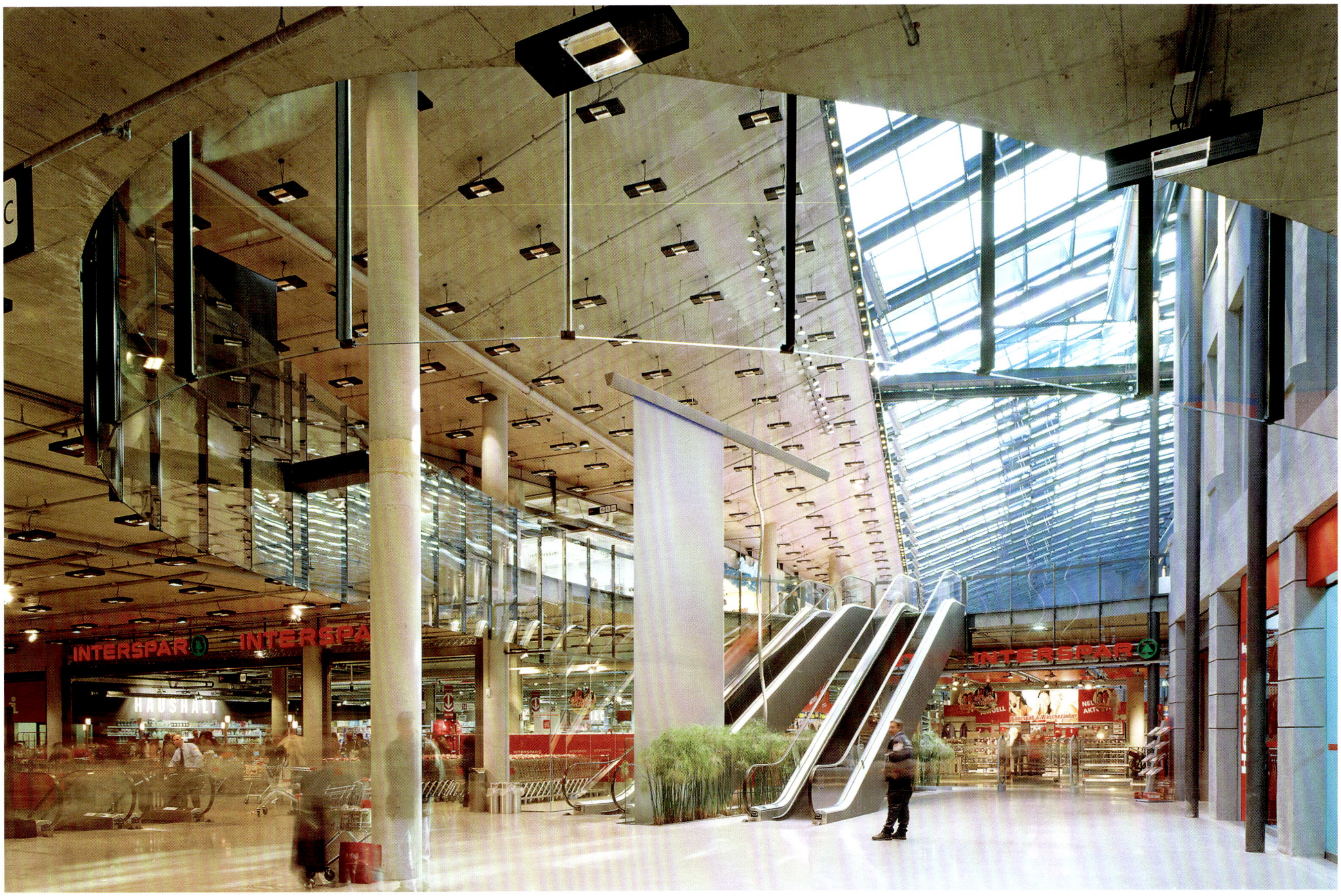
INTERSPAR
INTERSPAR
HAUSHALT
INTERSPAR

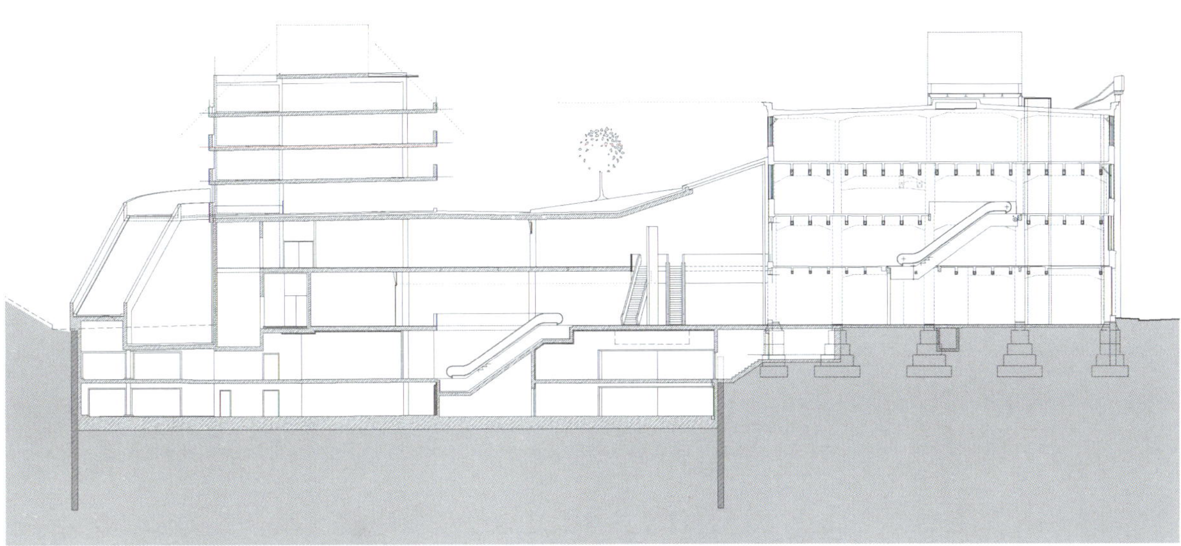

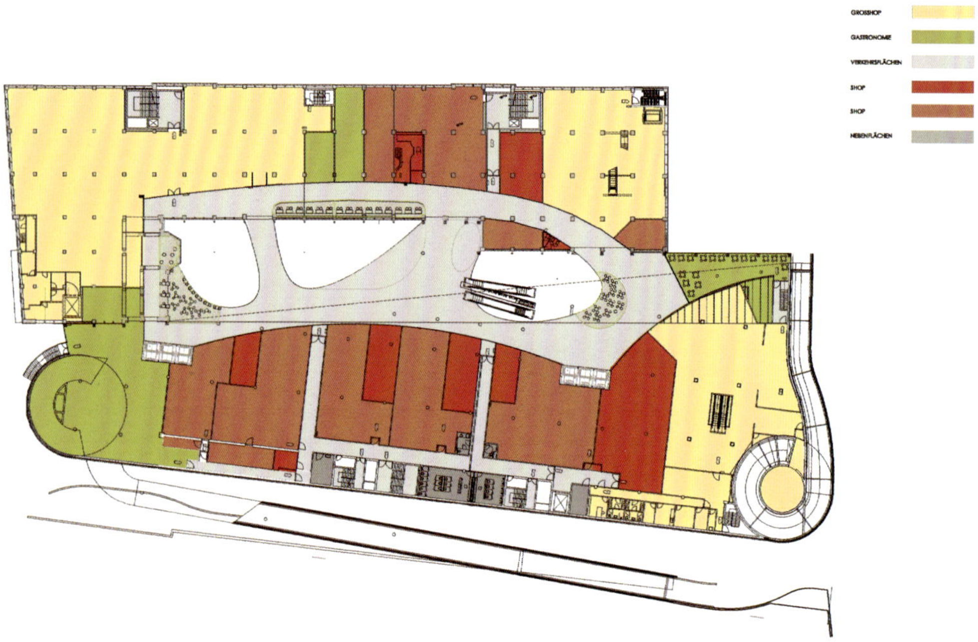

GROSSHOP
GASTRONOME
VERKEHRSFLÄCHEN
SHOP
SHOP
NEBENFLÄCHEN

GROSSHOP
VERKEHRSFLÄCHEN
SHOP
SHOP
NEBENFLÄCHEN
BÜROFLÄCHEN
PARKDECK

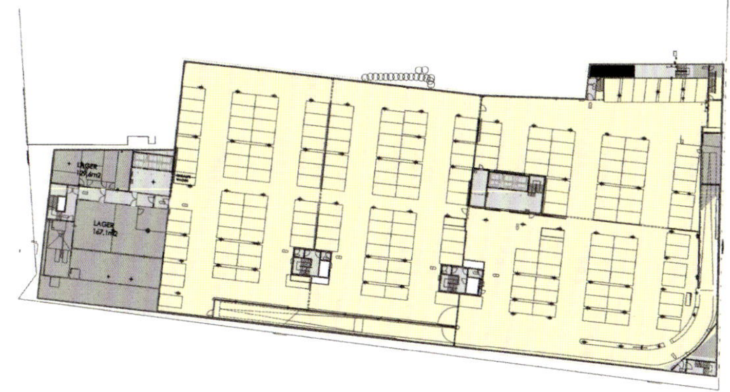
VERKEHRSFLÄCHEN
PARKDECK U1
NEBENFLÄCHEN

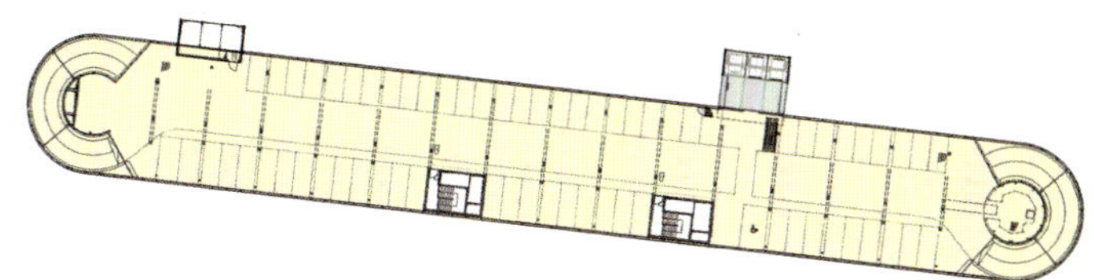
PARKDECK
DACHGESCHOSS

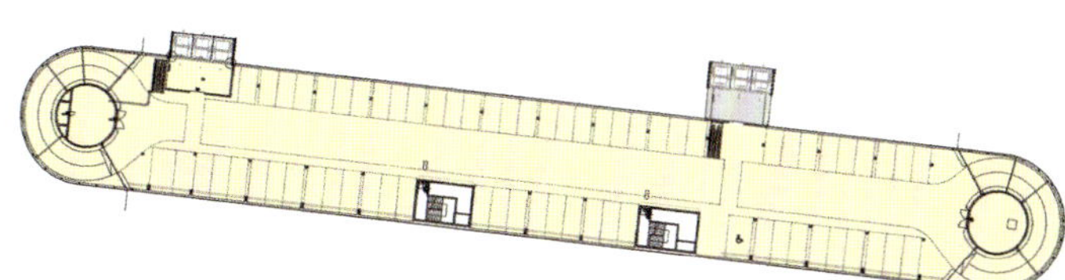

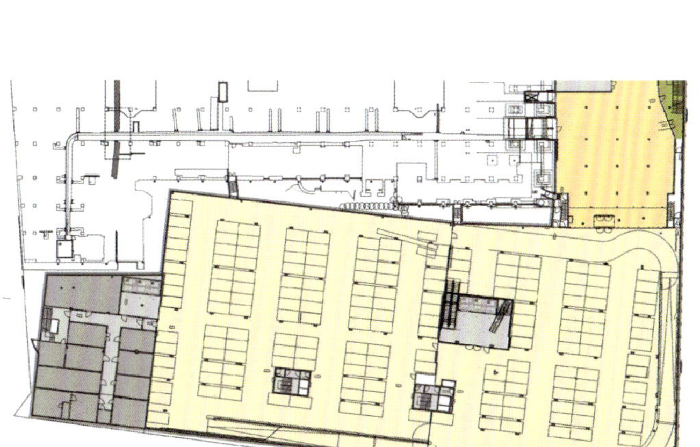
GROSSHOP
VERKEHRSFLÄCHEN
GRÜNFLÄCHEN U1
PARKDECK U1
NEBENFLÄCHEN

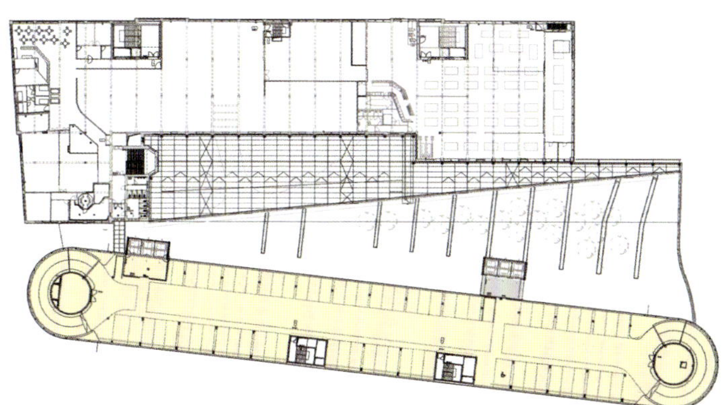

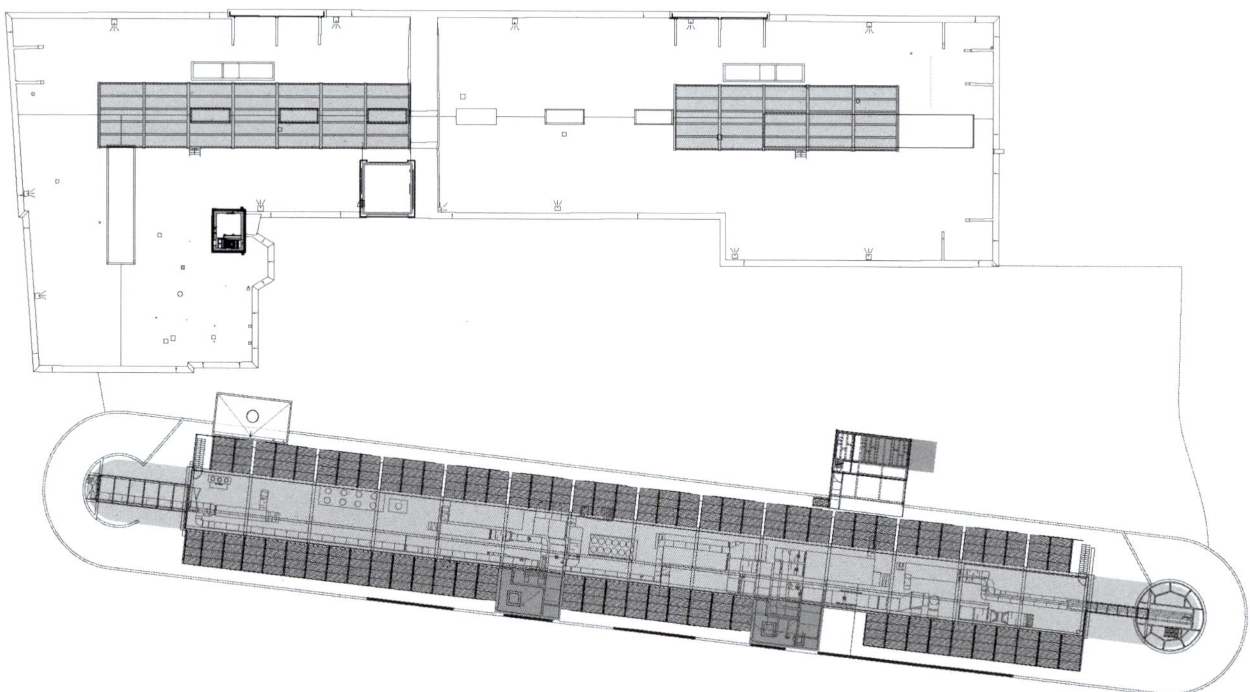

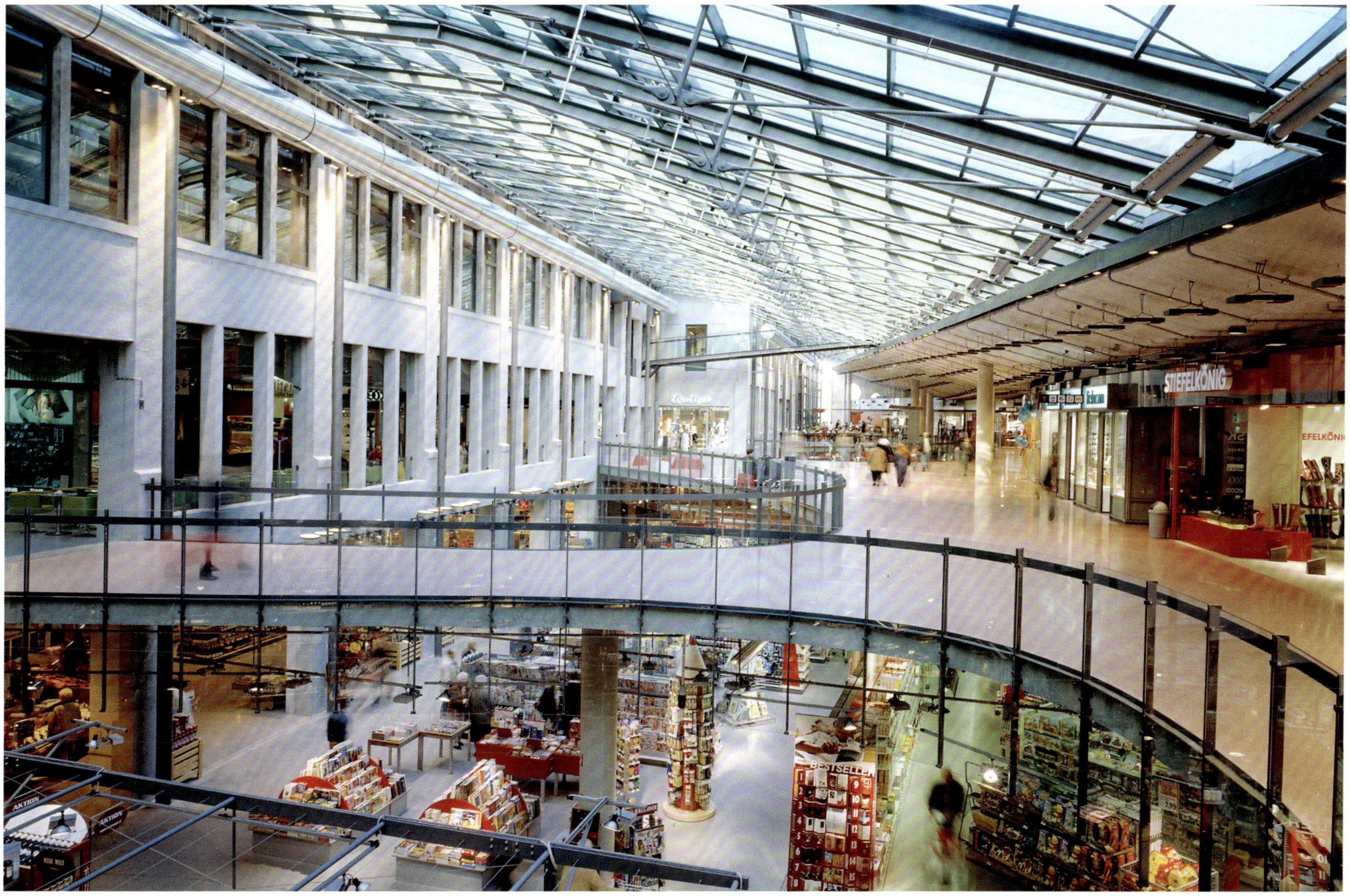
STIEFELKÖNIG

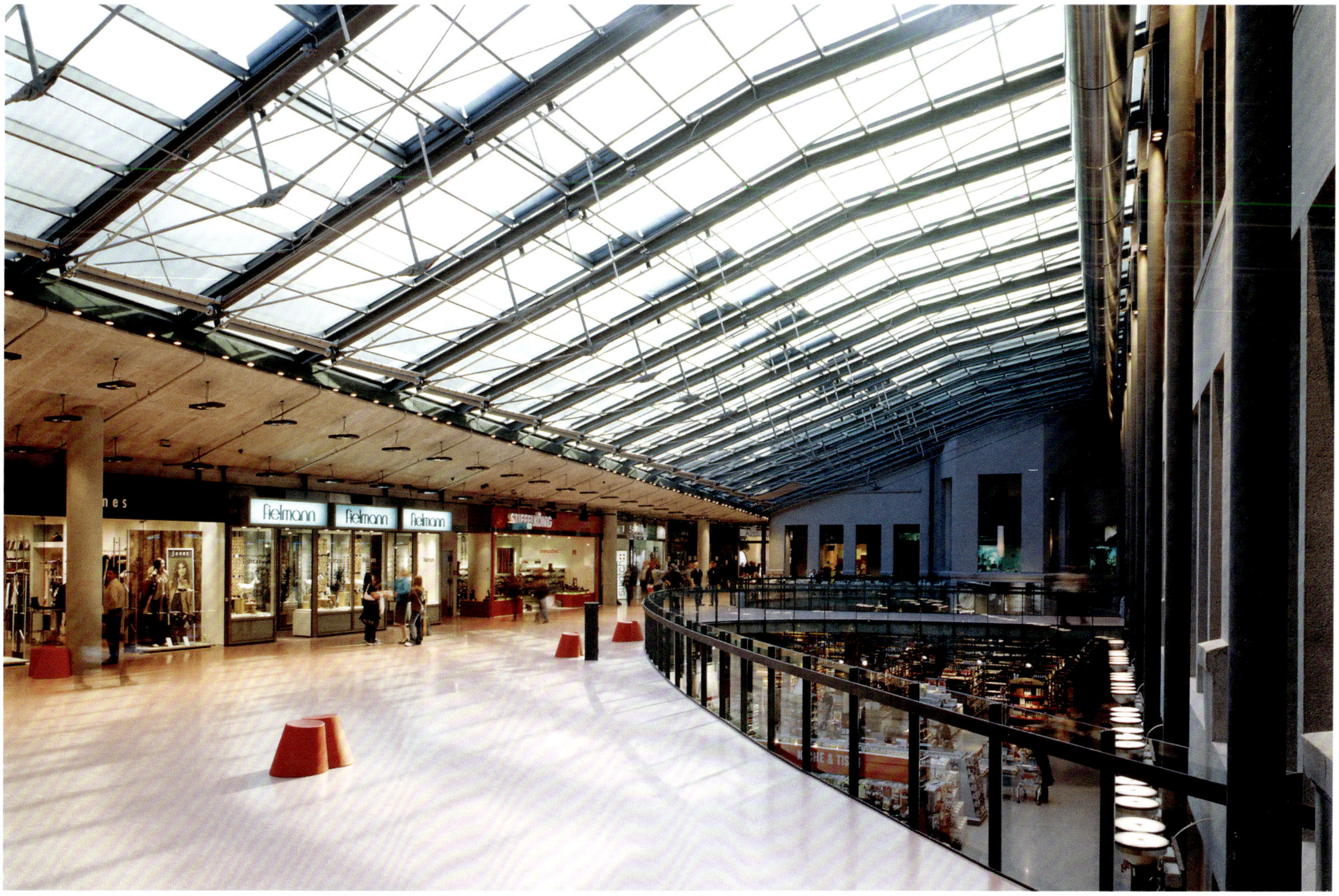
fielmann
fielmann
fielmann

# Jockey Plaza Boulevard
# 赛马广场大道

This project has been developed in two levels, recreating a vila with small buildings and creating a controlled urban experience. The idea of this proposal is to create a special and fully equipped meeting point as well as a place to interact in a small scale.That´s how the scheme is based on three articulating elements:

1)The main circular square, which is the most important space and thought as a big attraction. It is playful and its use is associated with entertainment.

2)The central axis, symbolic element that will become the icon of the mall.

3)The new technology applied to create a day - night experience by the use of LEDS, landscaping and façade treatments.

该项目有2层，重建了一个带有小型建筑的维拉港和创建一种控制城市的体验。这项提议的想法是建造一个特殊的和装备完善的交汇点，以及一个小规模的互动地方。这是该计划如何基于三个关联元素而进行的阐述：

1）主要的圆形广场作为一个极具吸引力的重要空间，它充满乐趣，它的使用与娱乐相关；

2）中轴，是象征性的元素，这将成为该商场的图标；

3）新技术应用，通过LED灯的使用、环境美化和外观处理营造了一个日夜体验。

**Company**
Arq Jose Orrego

**Designer**
Arq Jose Orrego

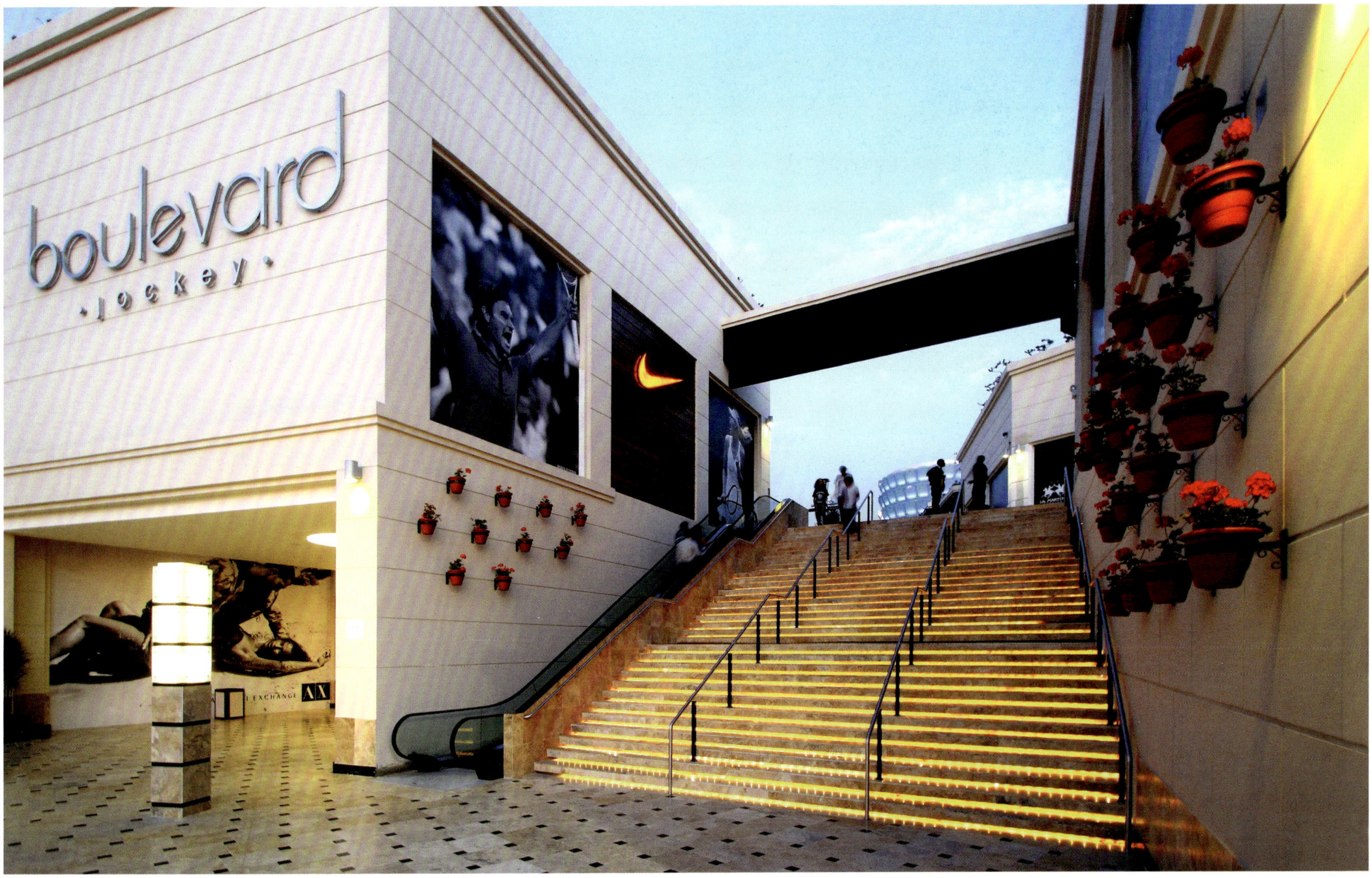
boulevard
jockey
EXCHANGE

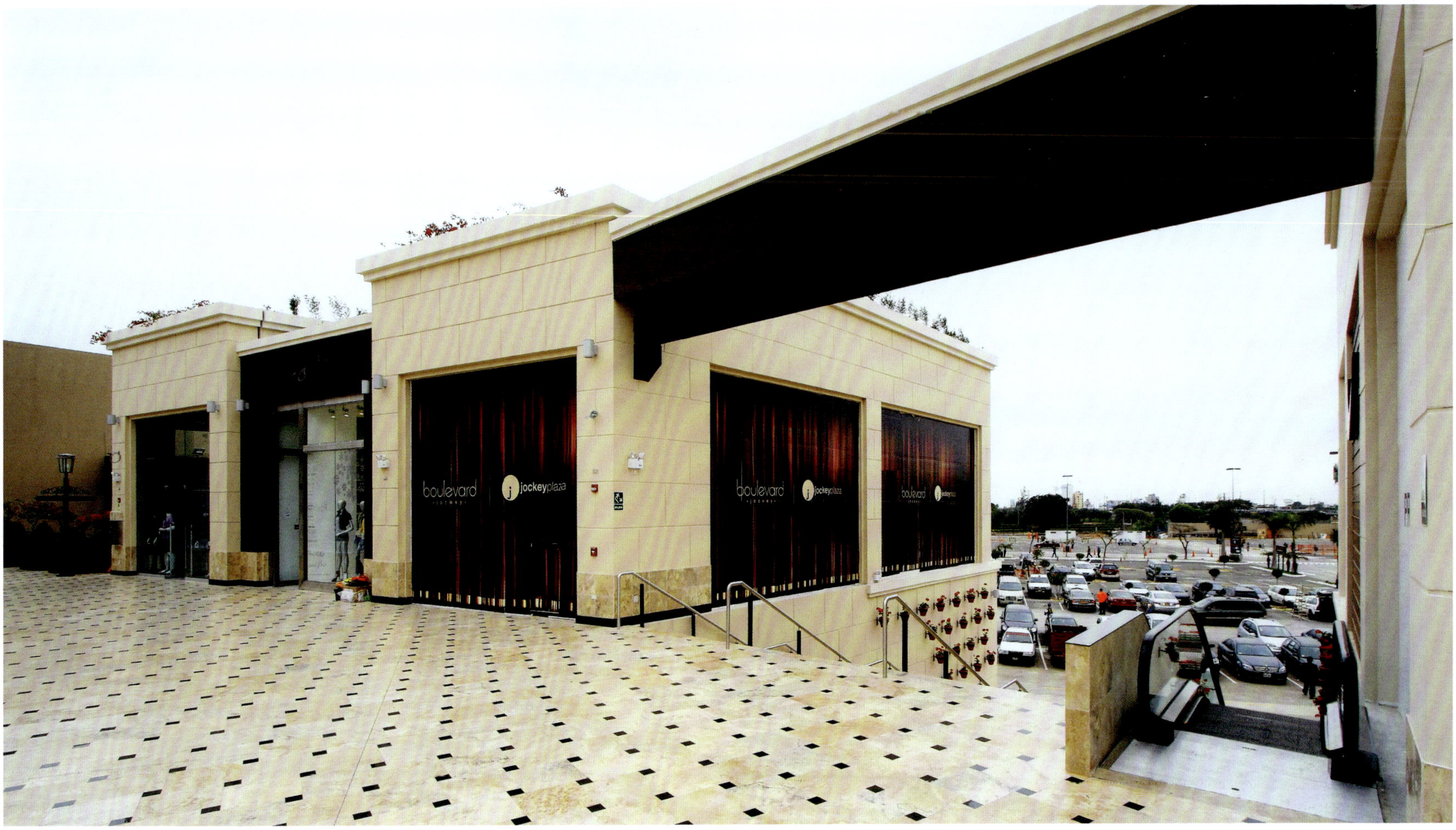
boulevard
jockeyplaza

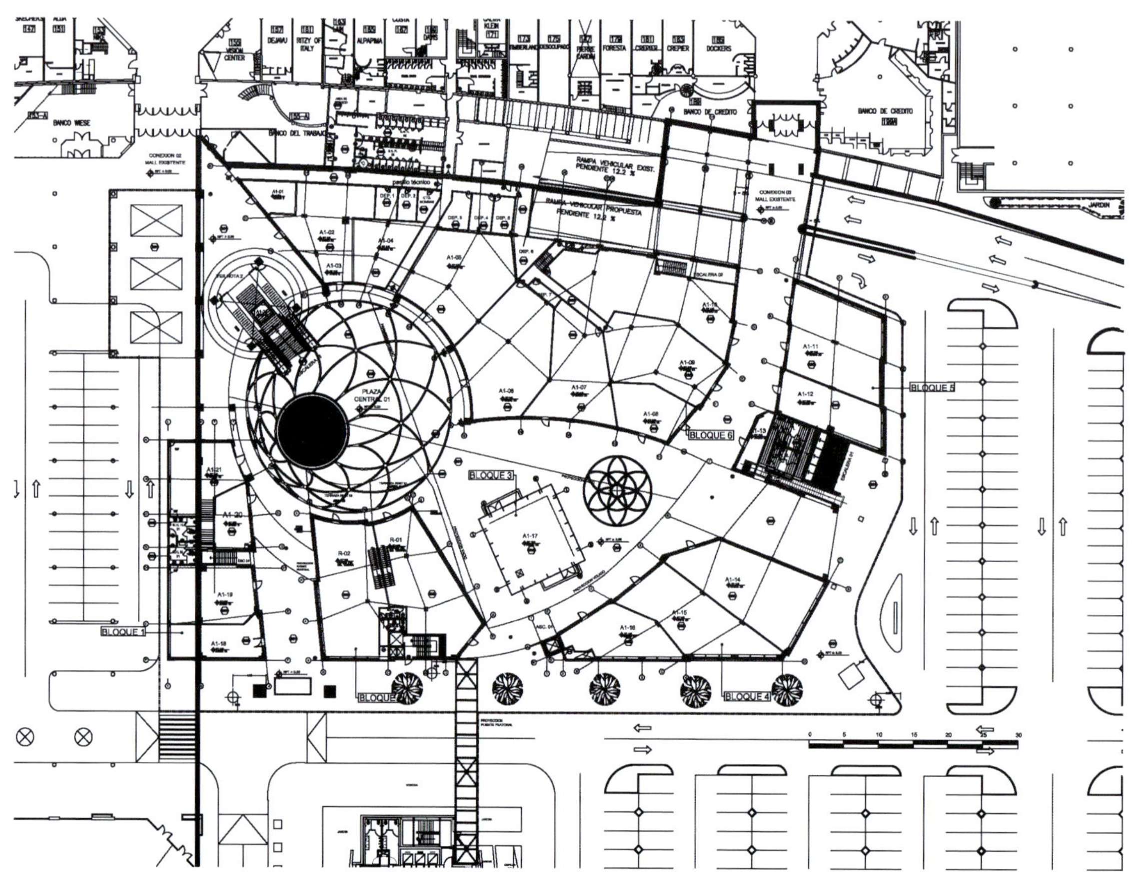
BANCO DE CREDITO
RAMPA VEHICULAR EXIST. PENDIENTE 12.2 %
RAMPA VEHICULAR PROPUESTA PENDIENTE 12.2 %
BLOQUE 5
BLOQUE 6
BLOQUE 3
BLOQUE 4
BLOQUE 1
DOCKERS
ALPARGA
BANCO WIESE

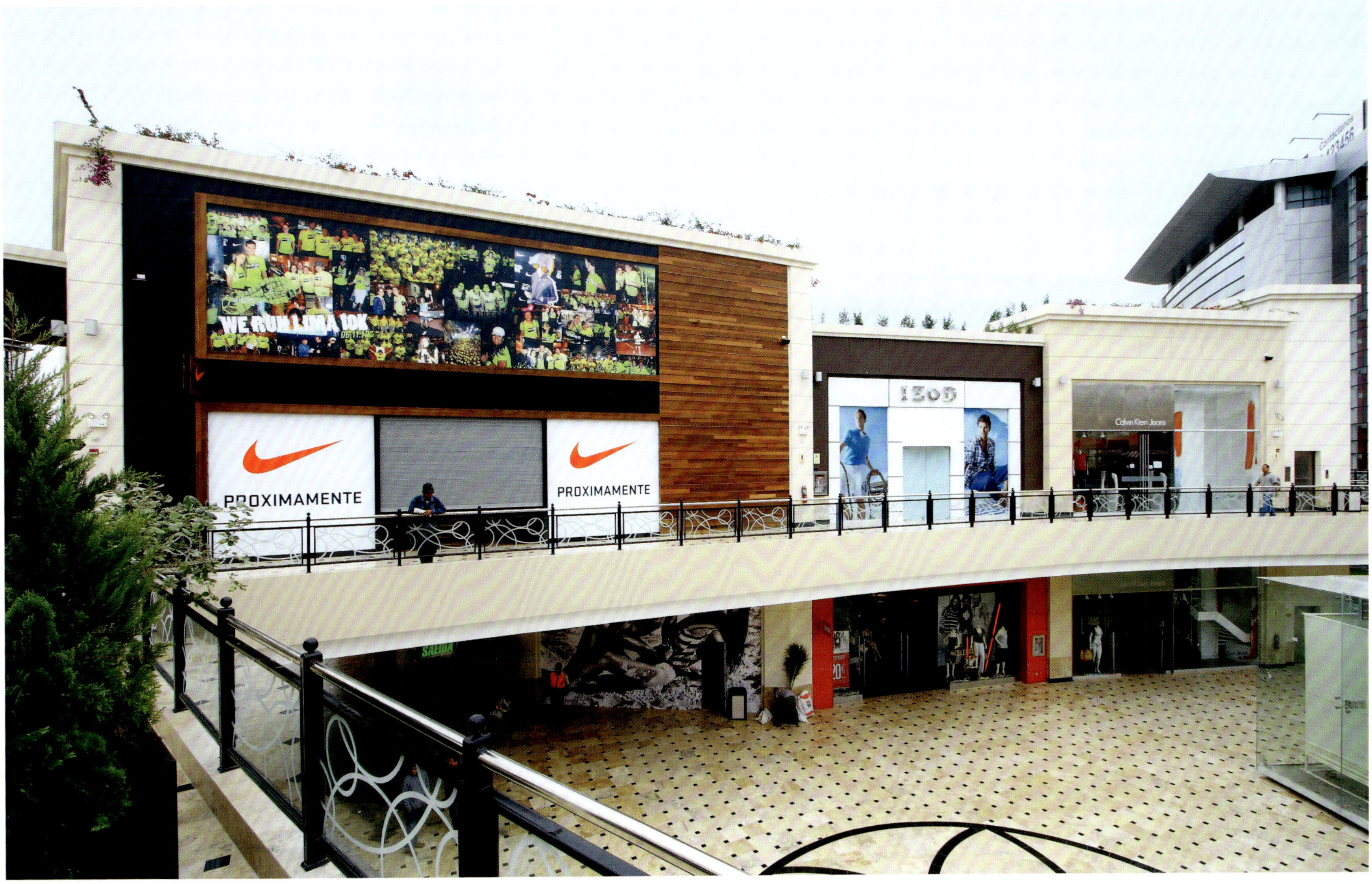
WE RUN LIMA 10K
PROXIMAMENTE
PROXIMAMENTE
IZOD
Calvin Klein Jeans
SALIDA

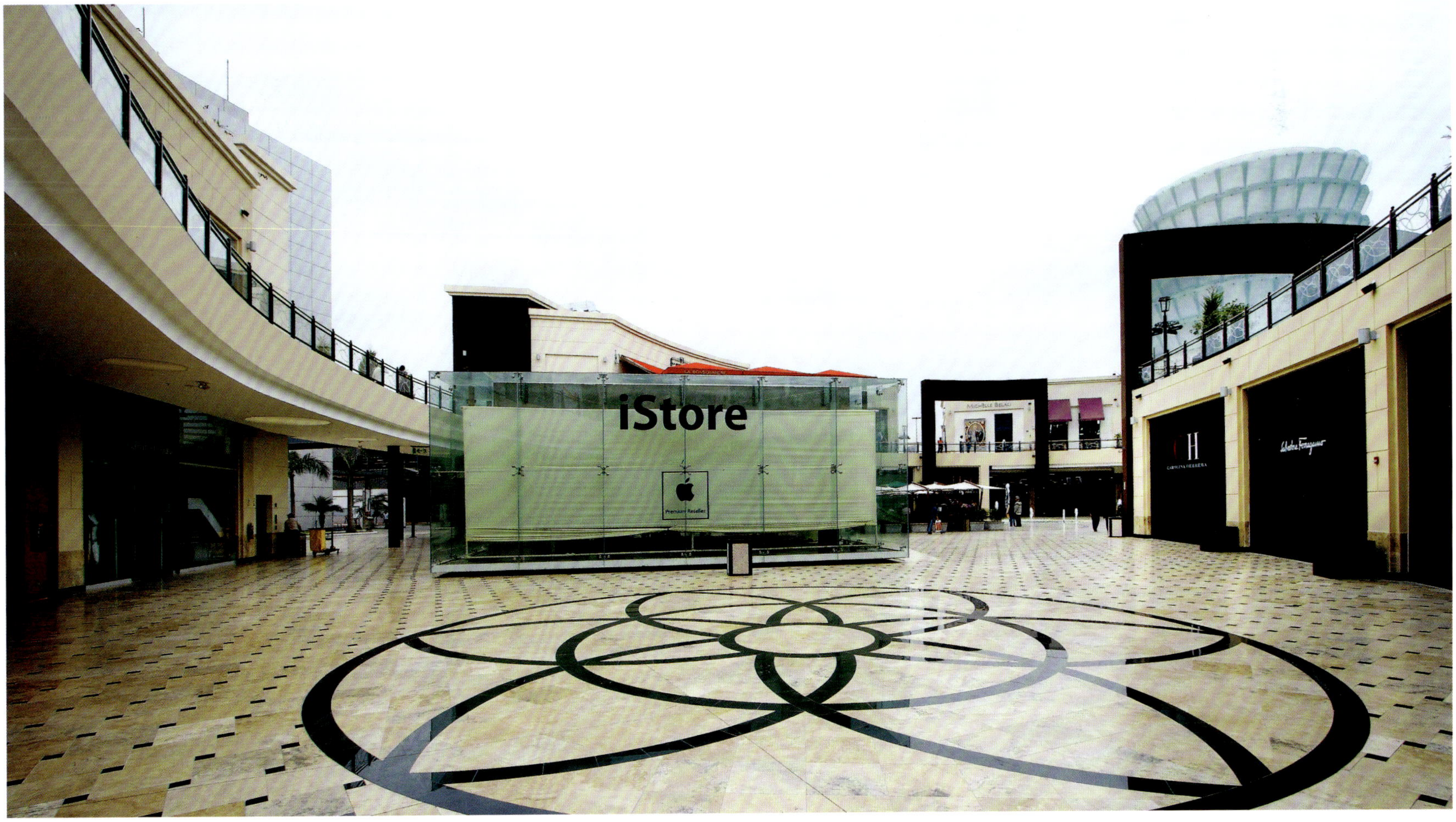
iStore

LA BONBONNIERE
LA BONBONNIERE
iStore
iStore
Bodega de la Trattoria

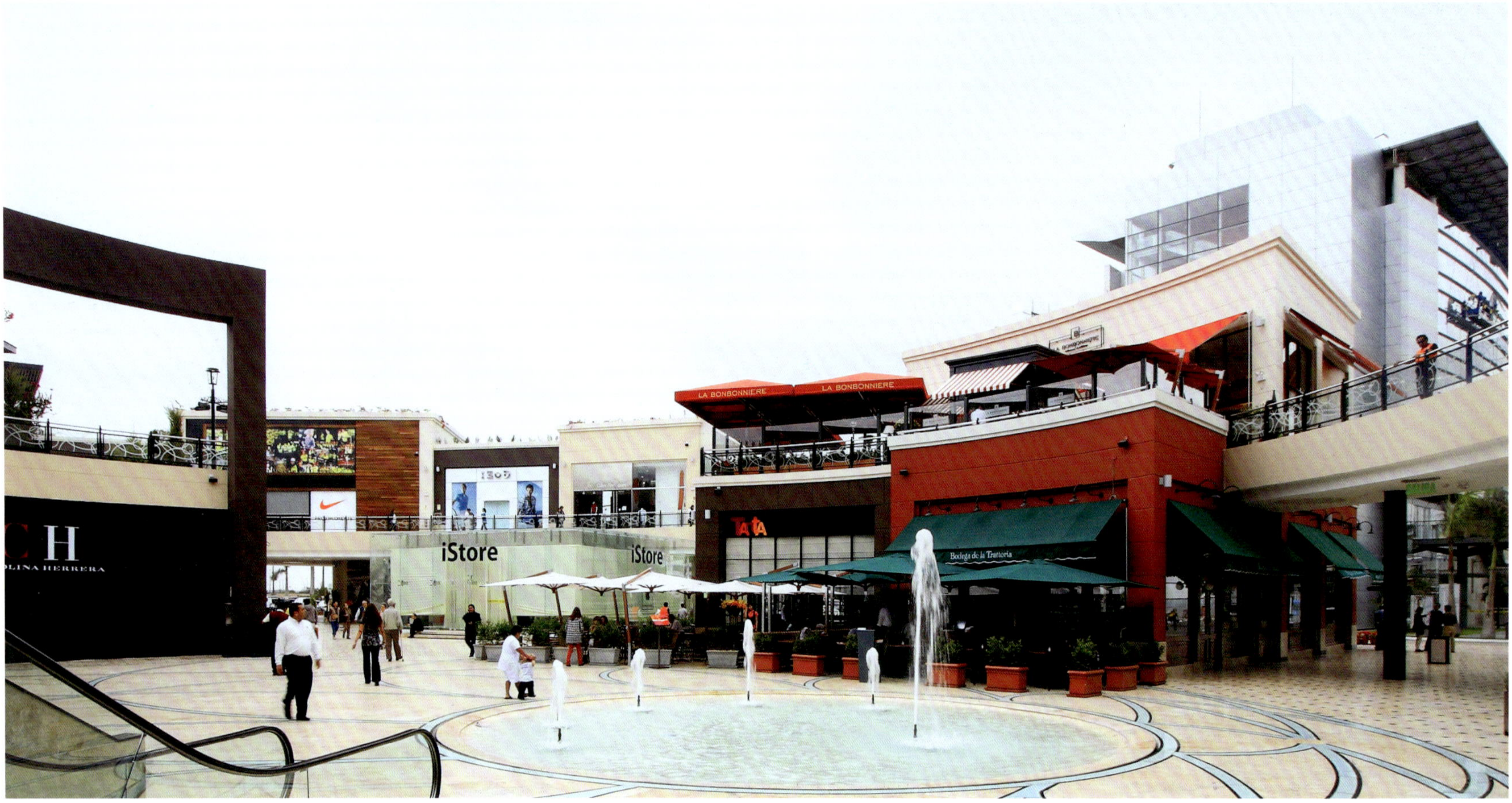

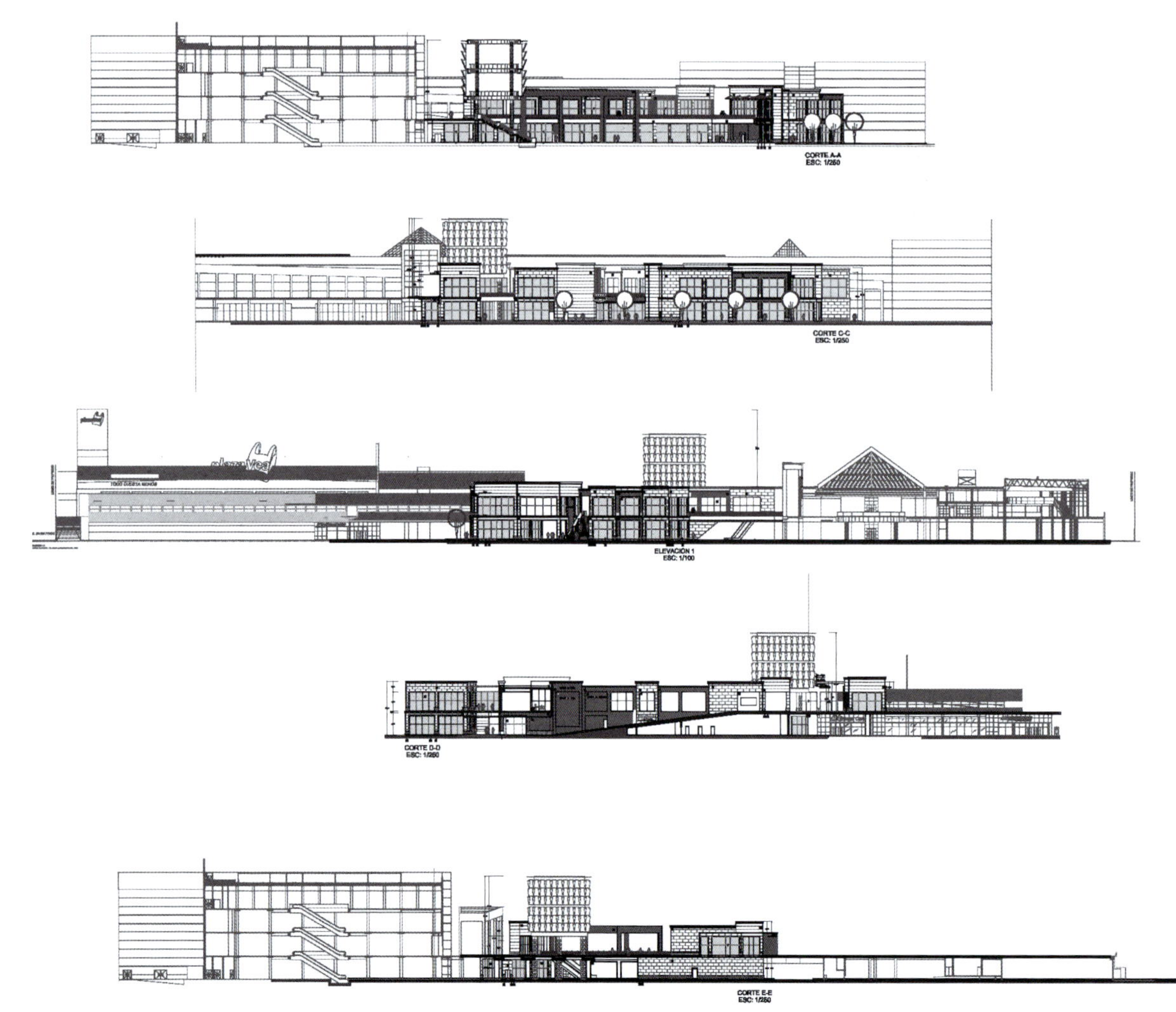

CORTE A-A
ESC: 1/250

CORTE C-C
ESC: 1/250

ELEVACION 1
ESC: 1/100

CORTE D-D
ESC: 1/250

CORTE E-E
ESC: 1/250

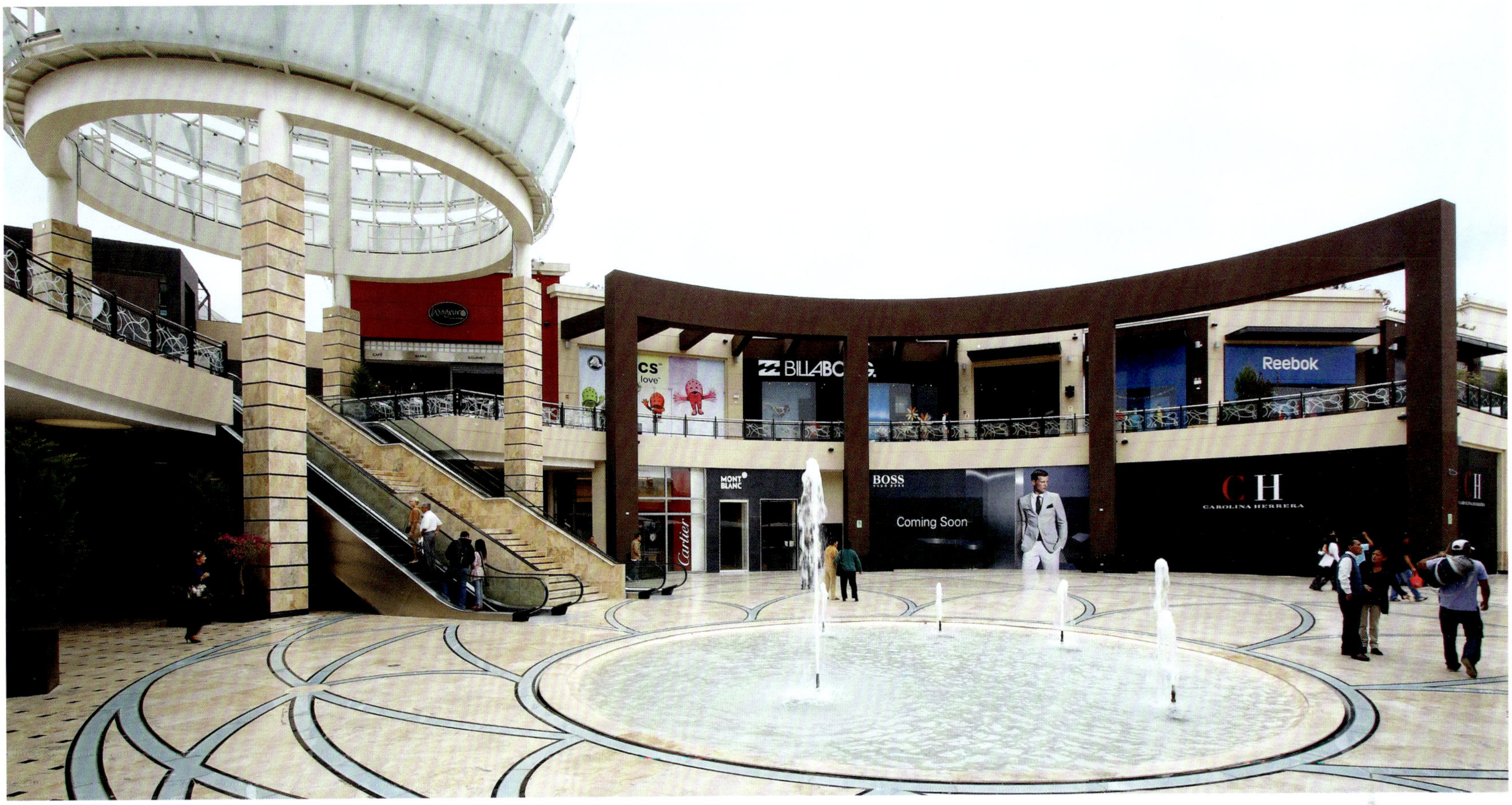
BILLABONG
Reebok
MONT BLANC
BOSS
Coming Soon
Cartier
CH
CAROLINA HERRERA

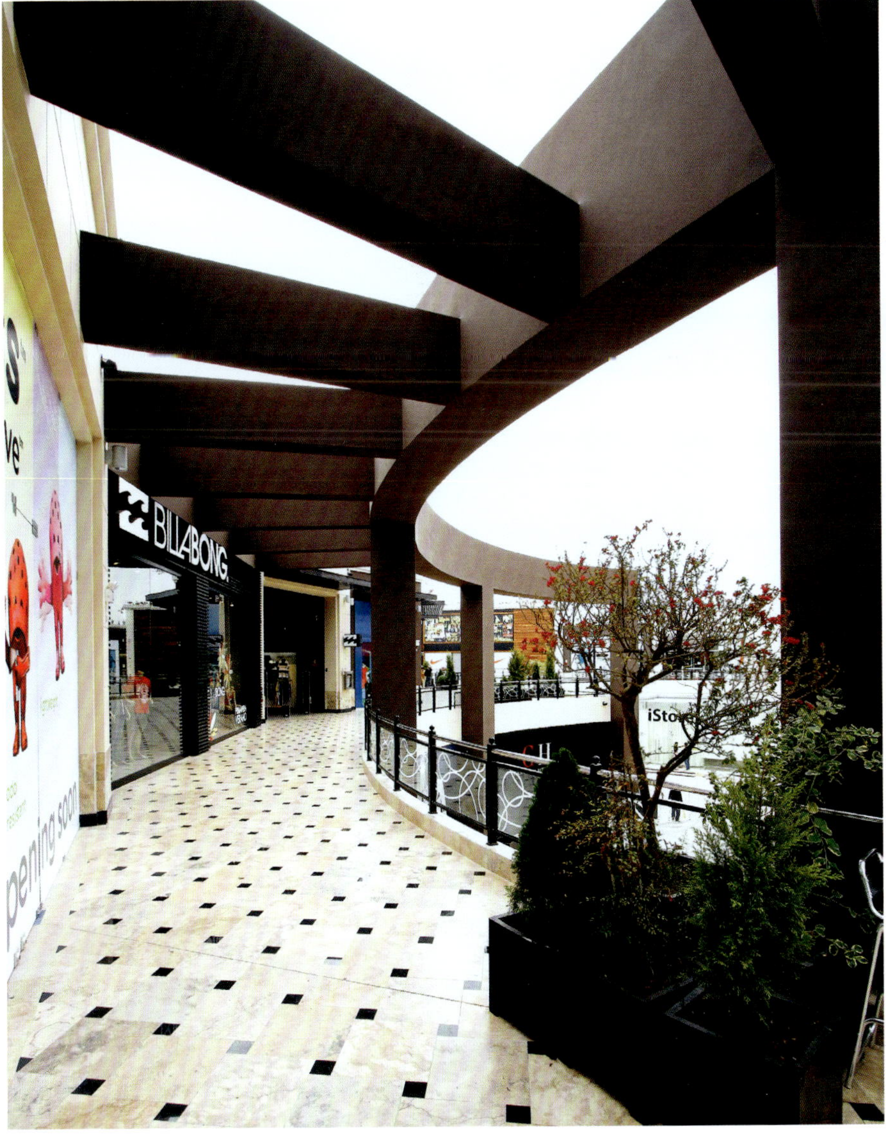

BILLABONG

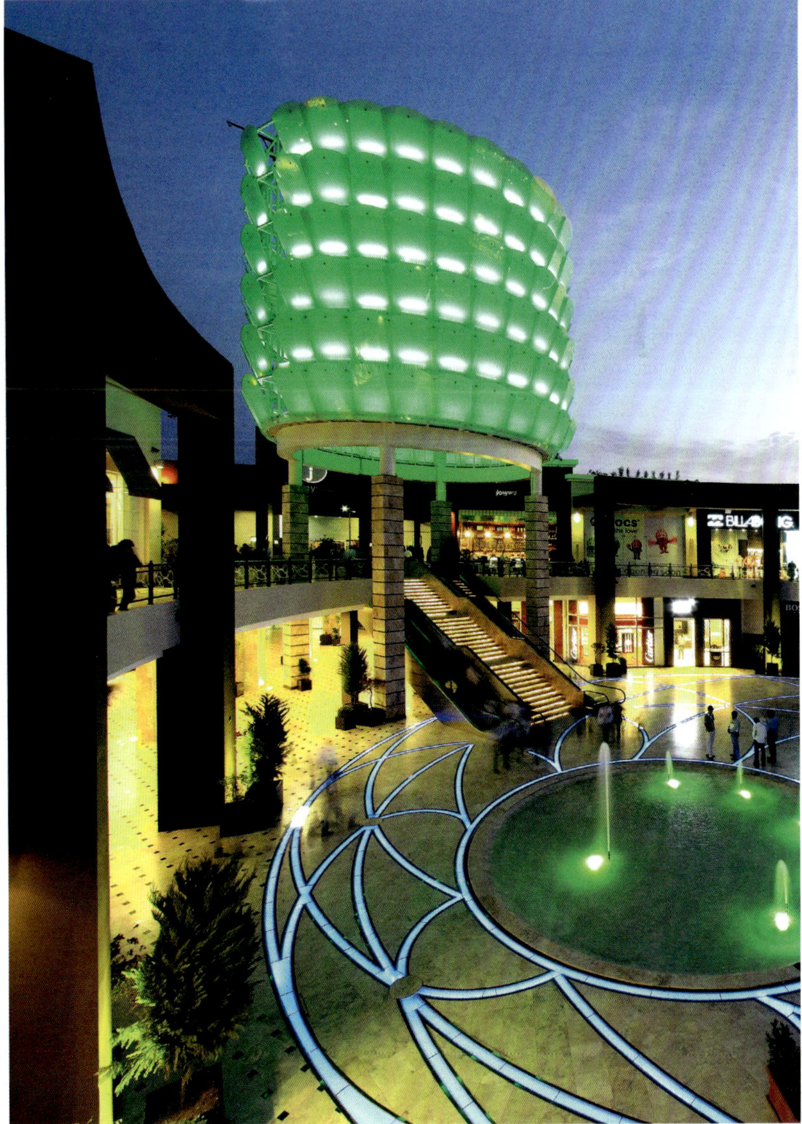

# Amstlvn

## Amstlvn购物中心

Shopping Centres owner, Unibail Rodamco, wants to be a host to their clients; with this vision they developed the so - called "Welcome Attitude Program". By providing services and comfort - zones they aim at making the visitors feel welcome and informed.

Studio Linse integrated Unibail Rodamco's Welcome Attitude in shopping centre Amstelveen. In the design more cohesion has been created between the different parts of the shopping centre by using specific colours and connecting interior aspects, like special lampshades. The colors used are green and warm grey. The greenery provides a fresh and youthful appearance, where the warm grey gives a luxurious look and feel.

The customers' journey may start at the website, at the parking area or at the entrances. Customers are welcomed by clear wayfinding information supported by inviting visuals. All these elements are complementary, making the "Welcome Attitude" vision recognised by the customers in all areas. Besides this overall plan of upgrading the shopping centre, Unibail Rodamco is continuously busy with the positioning of the centre, as the most comfortable shopping centre of the greater Amsterdam area.

**Location**
Amstelveen

**Arca**
78, 200 sq.m.

**Main Materials**
Painted walls, vinyl visuals, backlit opale glass, white painted steel, Wenge wood, etc.

**Photographer**
Michel Claus

**Company**
Studio Linse

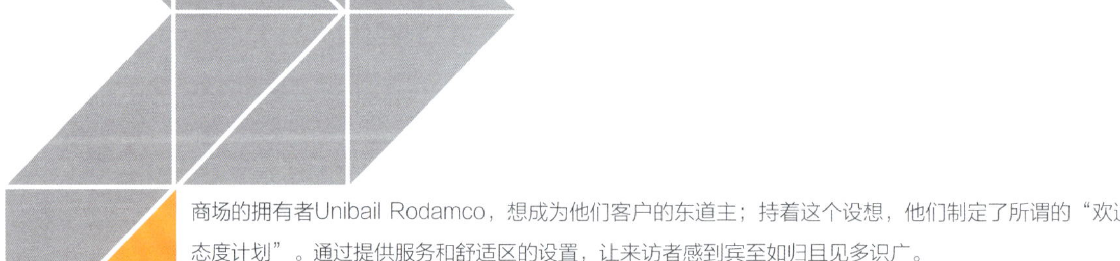

商场的拥有者Unibail Rodamco，想成为他们客户的东道主；持着这个设想，他们制定了所谓的“欢迎态度计划”。通过提供服务和舒适区的设置，让来访者感到宾至如归且见多识广。

Studio Linse在阿姆斯多芬购物中心结合了Unibail Rodamco的迎宾态度。在本案设计中，通过使用特定的颜色并且连接了内饰各方面，如特殊的灯罩，使得在购物中心不同部分之间创造了更多凝聚力。颜色方面使用的是绿色和暖灰色，绿化提供了一种清新和年轻的外观，暖灰色赋予其奢华的外观和感觉。

客户可能会在网站上、在停车区或在入口处开始他们的旅程。迎接客户的是通过邀请视像提供了清晰的寻路信息，所有这些因素相辅相成，让“欢迎态度”的愿景得到各区域客户的认可。除了这个改善购物中心的总体规划，Unibail Rodamco一直忙于该中心作为大阿姆斯特丹地区最舒适的购物中心的定位。

amstlvn
amstlvn
de Bijenkorf
amstlvn

info
MEYER & MEYER
- tailors -

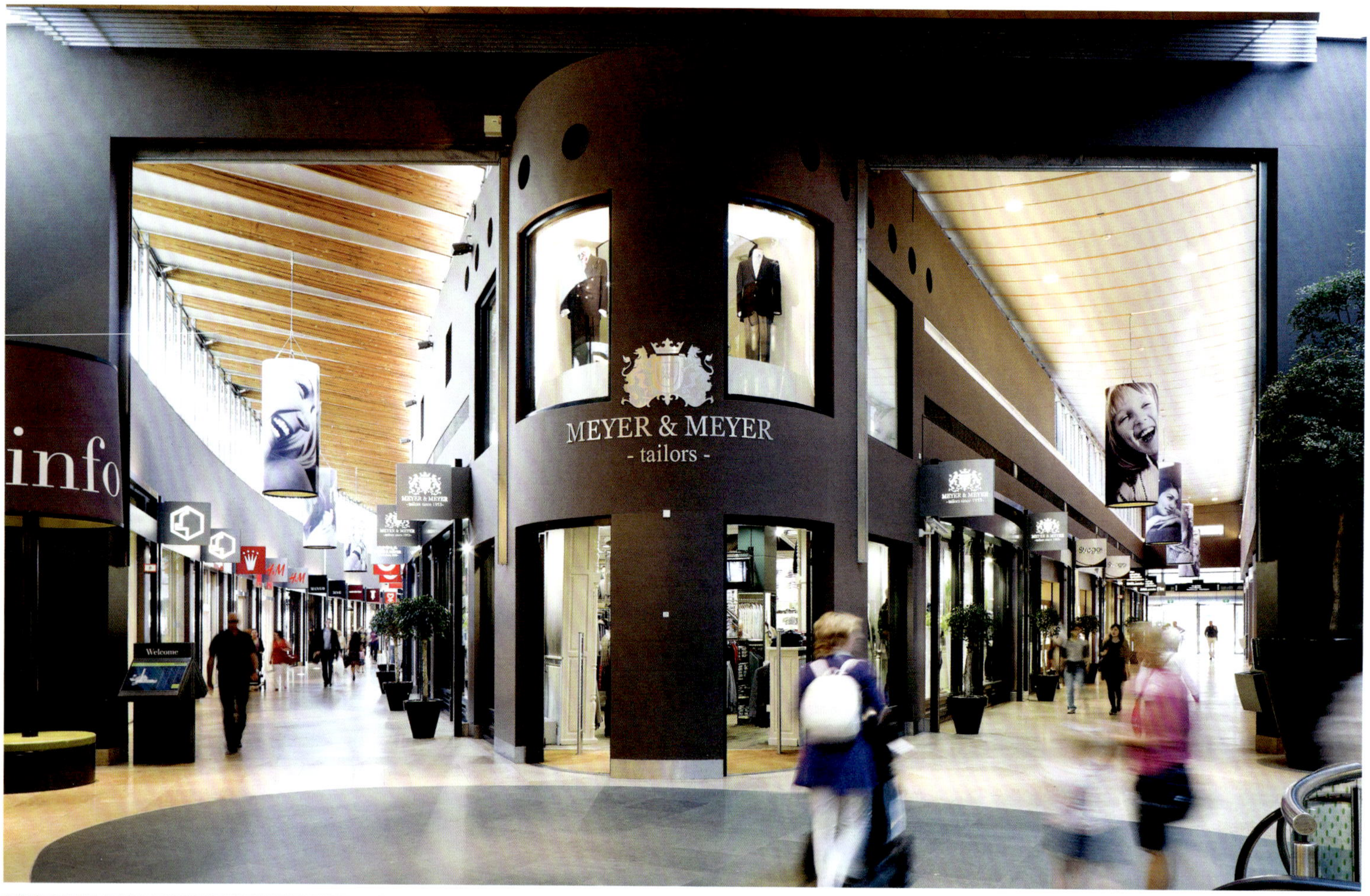
info
MEYER & MEYER
- tailors -
Welcome

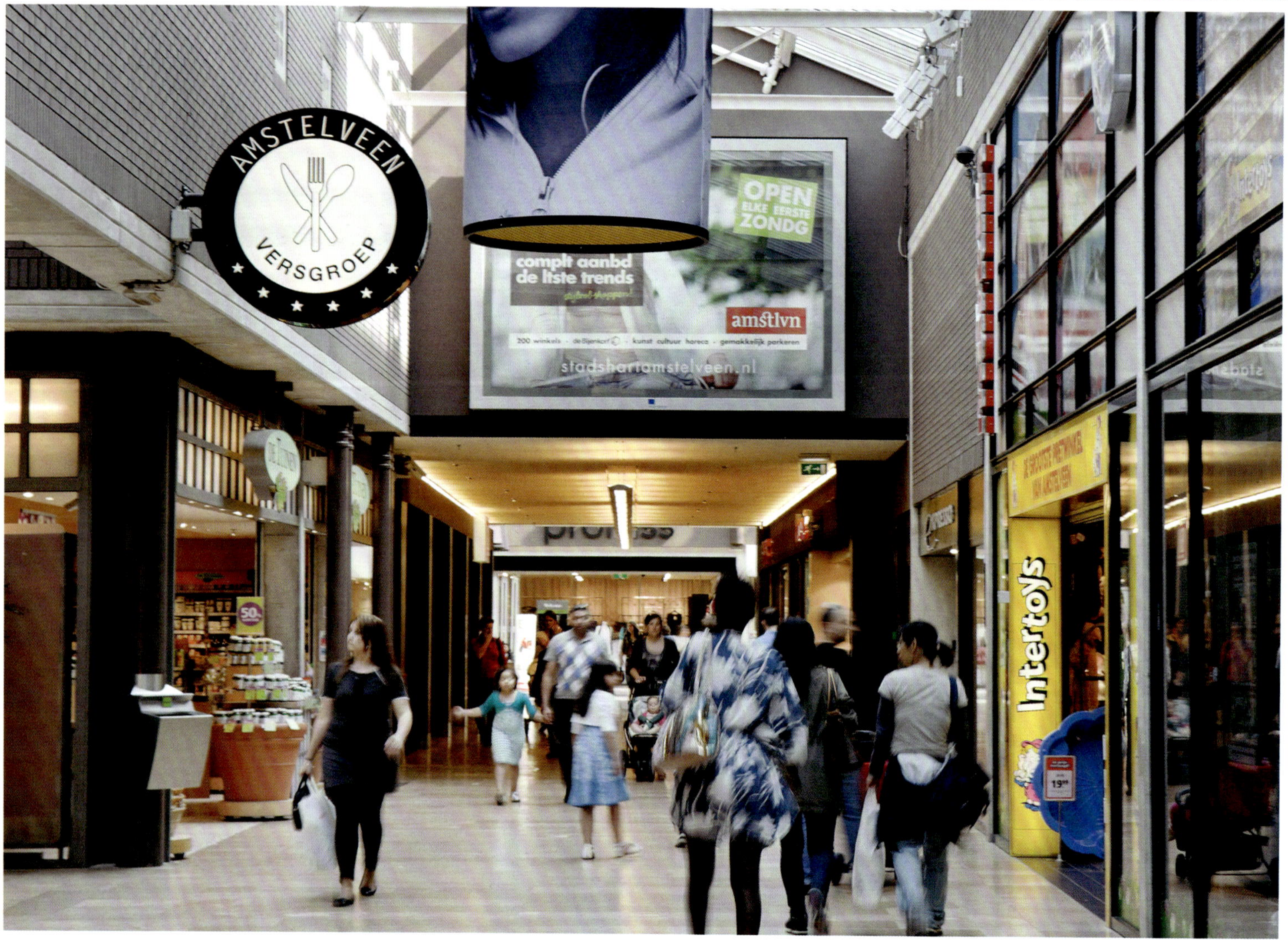
AMSTELVEEN
VERSGROEP
OPEN
ELKE EERSTE
ZONDG
complt aanbd
de liste trends
amstlvn
stadshartamstelveen.nl
Intertoys

amstlvn
amstlvn
PEARLE
amstlv
MEYER & MEYER
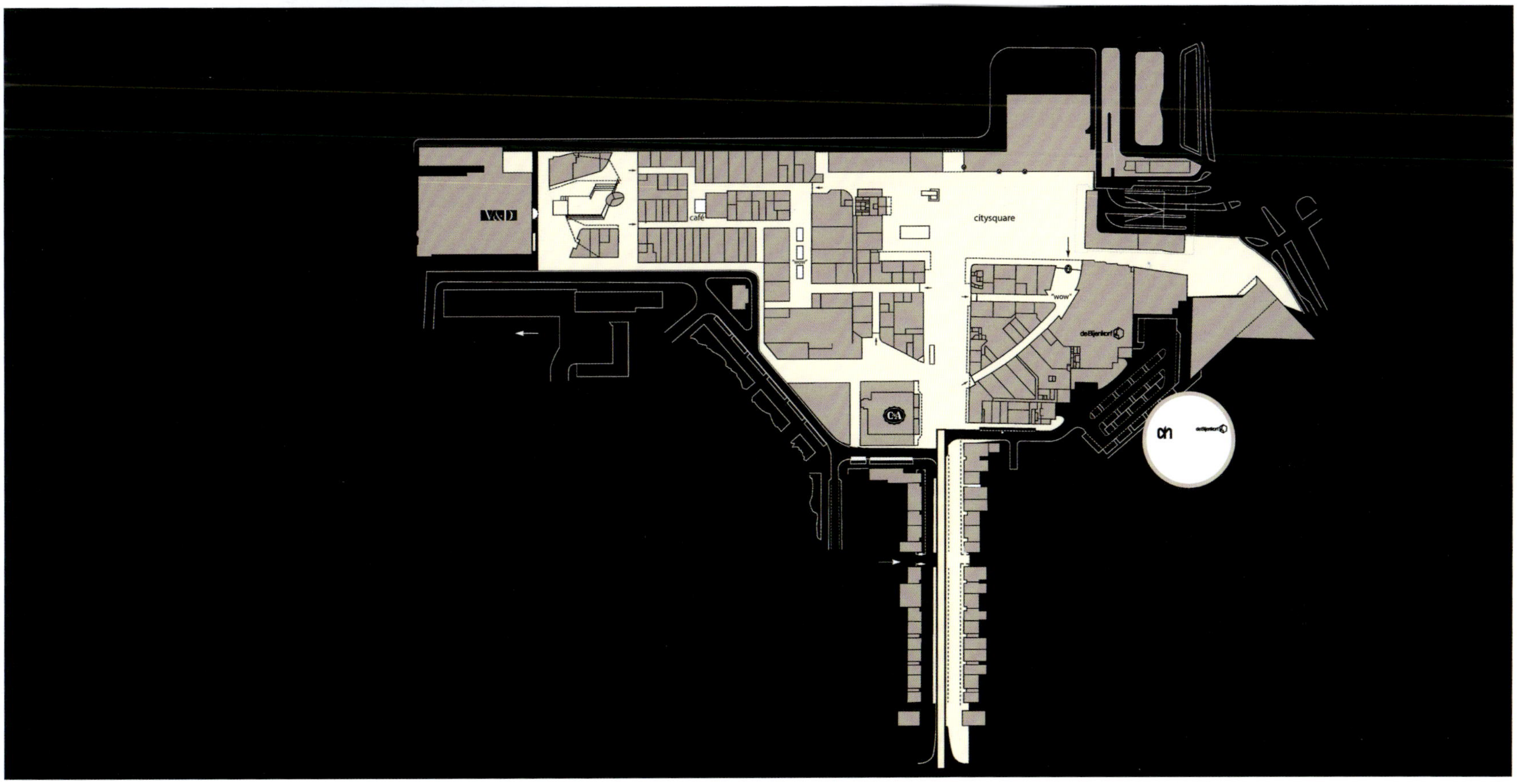
V&D
café
citysquare
"wow"
C&A

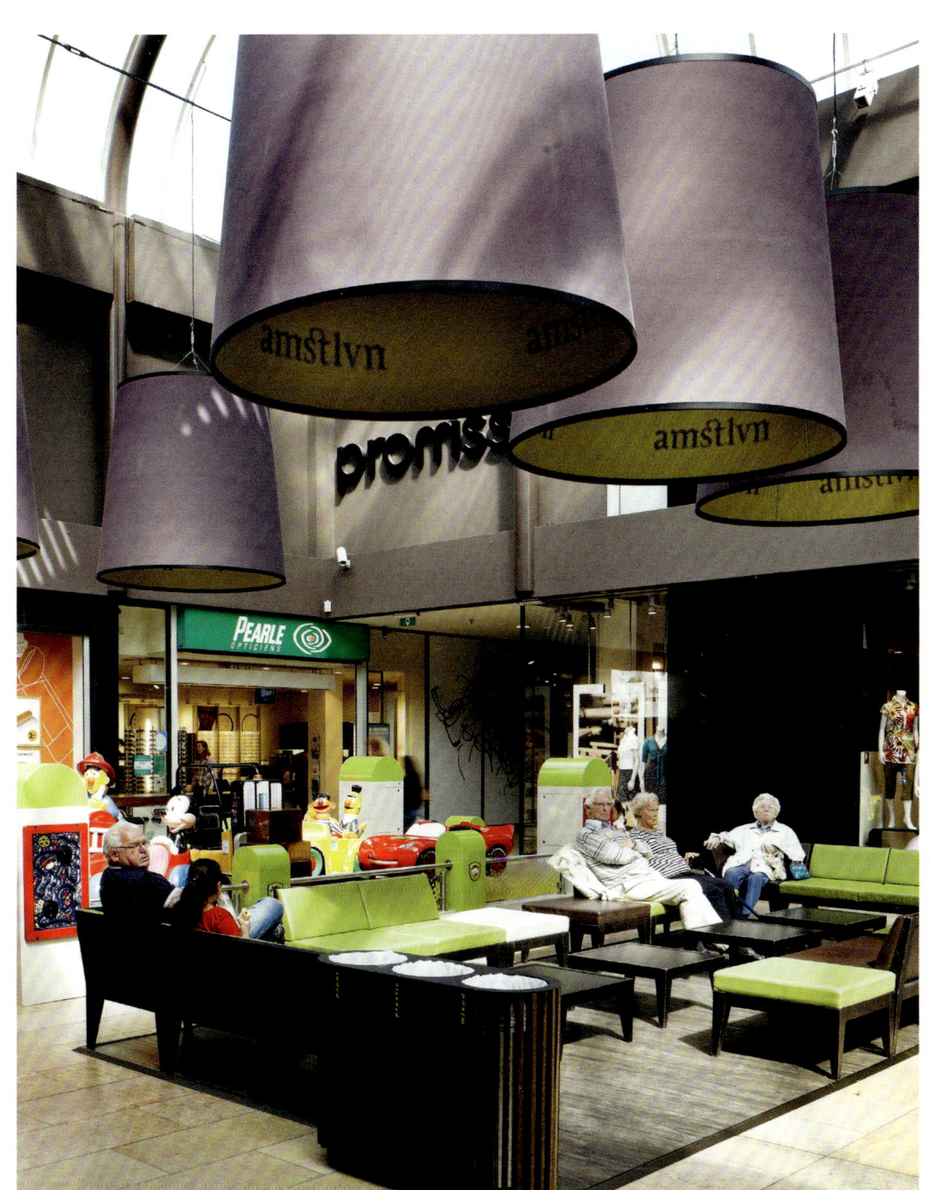

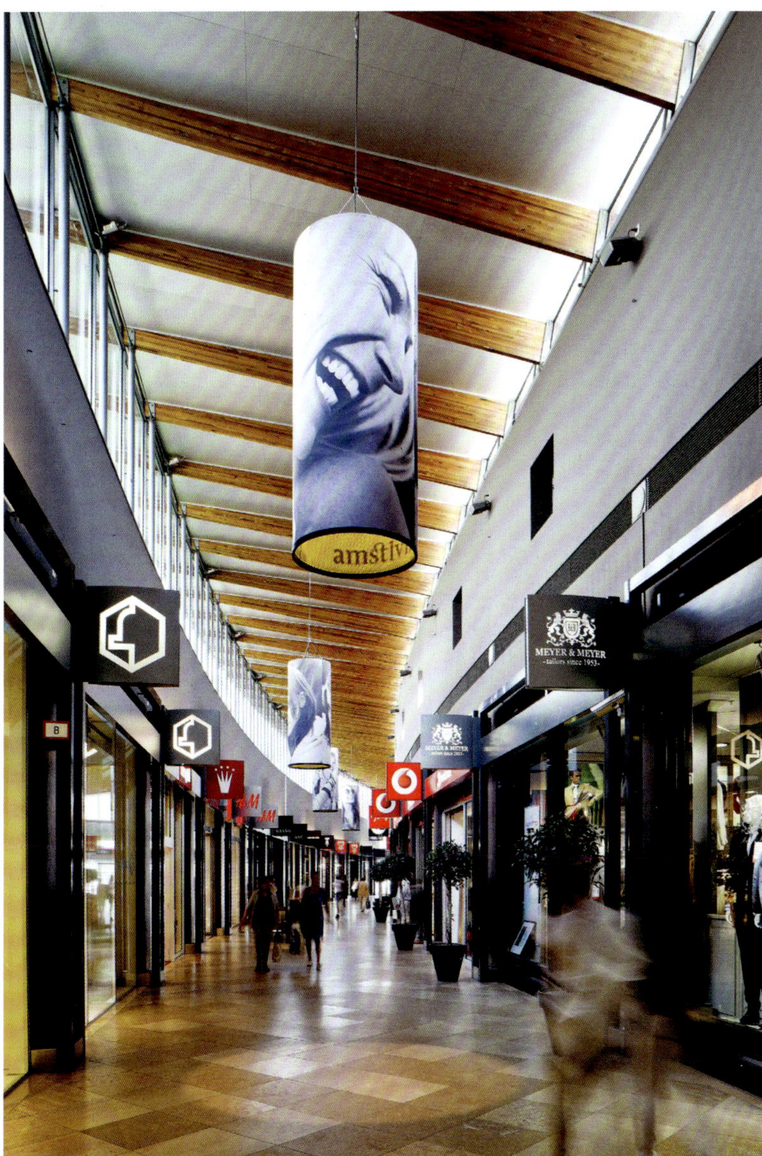

ZARA
ZARA

MOND DESIGN
EGBERTS Café
Douglas

BLOKKER
Douwe Egberts Café
iso

Hergo

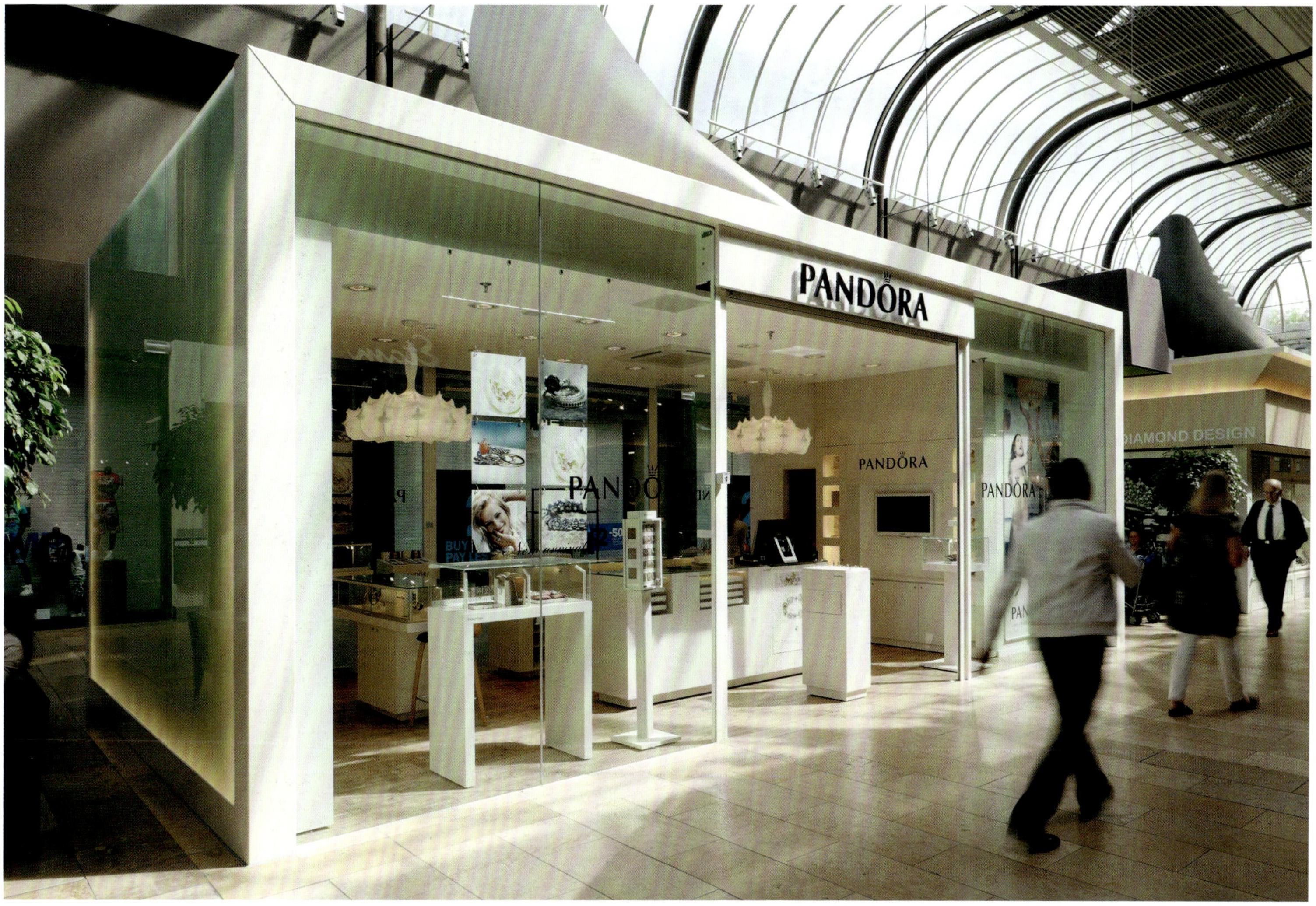
PANDORA
DIAMOND DESIGN

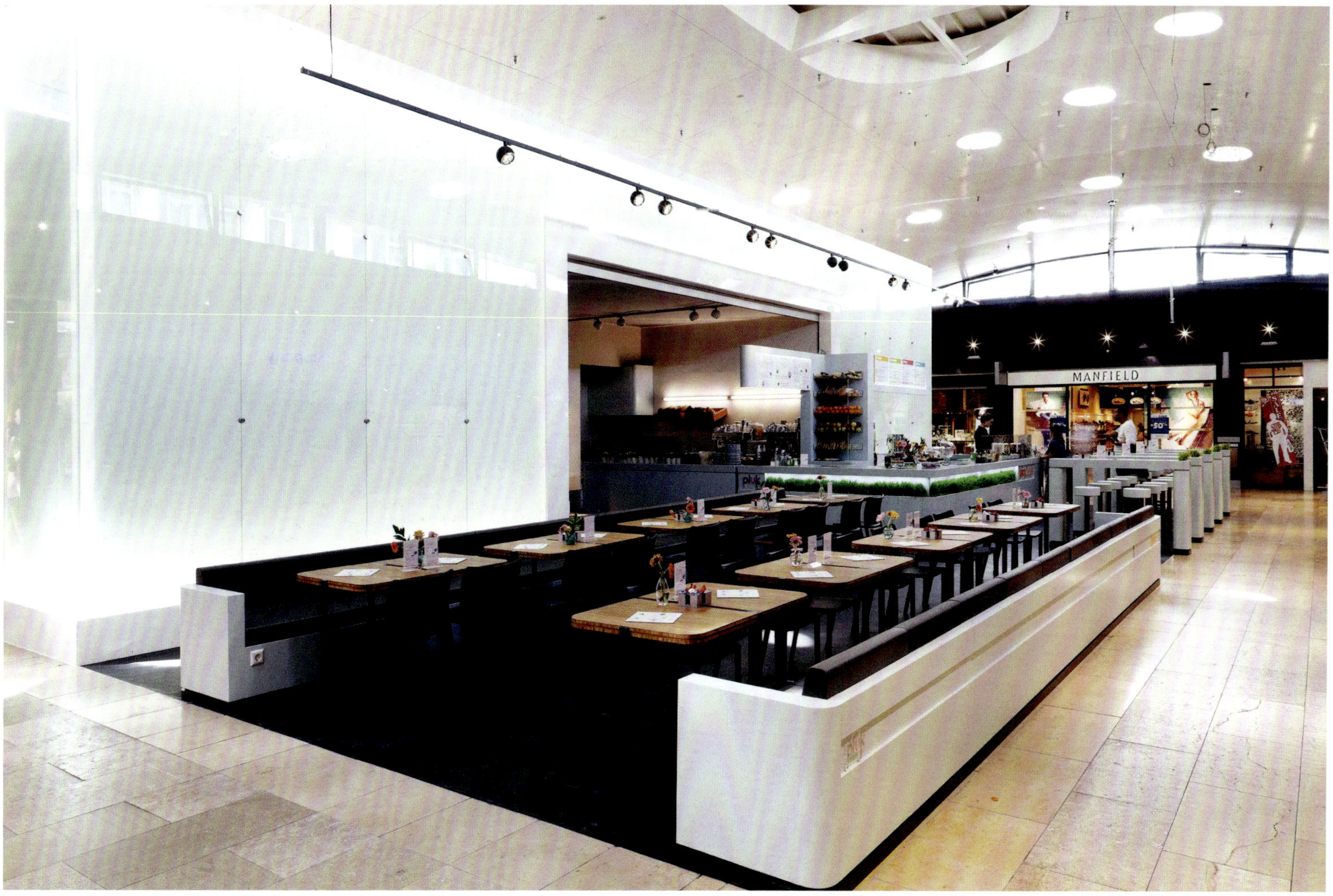
MANFIELD

# Forum Kapadokya Shopping Center

## 卡帕多西亚Forum购物中心

ERA performed the concept, design and execution phases of the project. The building is located at the center of Kapadokya region, with just 3 km of distance to Nevsehir city center and 80 km of distance to Kayseri. Forum Kapadokya project consists of a hypermarket, cinemas, shops, cafeterias, restaurants, approximately 25,000 sqm of leasable area and a car parking area with a capacity of 800 lots. Forum Kapadokya's concept is based on the modern interpretation of local architecture, natural formations and materials. The concept of the building is constituted with the inspiration of the troglodyte architecture which is the main character of the Kapadokya region. The design interprets and reflects the atmosphere of the spaces, material patterns and concepts like hollowness, indentation, carving specific to the region's morphology. Additionally, modern and simple forms composing the natural characteristic of the environment are considered as the main influential aspects of the conceptual design. The building's interior space is composed of several courts combined to each other with horizontal and vertical planes and streets on different levels. The intense use of local natural stone in the interior is as much as at the exterior; structural elements such as steel, glass and white walls are the fundamentals of the building. By the calm and warm effect of the local natural stone, the building differs greatly from the other shopping centers.

**Location**
Nevsehir, Turkey

**Area**
26,650 sq.m.

**Company**
ERA PLANNING ARCHITECTURE CONSULTING CO. LTD.

**Designer**
ERA PLANNING ARCHITECTURE CONSULTING CO. LTD.

ERA完成了这些项目的概念、设计和执行阶段。该建筑位于卡帕多西亚地区的中心，距离内夫谢希尔市中心仅3km，离开塞利有80km。本项目包括1个大卖场、一些电影院、商店、自助餐厅和饭店，可出租面积约25,000m$^2$，停车场能容下800个车位。本项目的概念是基于对当地建筑、自然形态和材料的现代诠释。建筑的概念由穴居建筑的灵感组成，这是卡帕多西亚地区的主要特色。其设计诠释并反映了空间的氛围、材料的形态和概念，比如针对该地区形态学的空旷、凹陷和雕刻。此外，构成环境自然特点的现代和简单的形状被看作方案设计的主要影响方面。该建筑的室内空间由几个相互结合的庭院组成，这些庭院带有水平和垂直的平面和不同层次的街道。室内和外观都一样大量使用当地天然石材。结构元素，如钢、玻璃和白色墙壁是该建设的基本材料。当地天然石材平静和温馨的效果使本建筑和其他购物中心的建筑大不相同。

forum
KAPADOKYA

forum

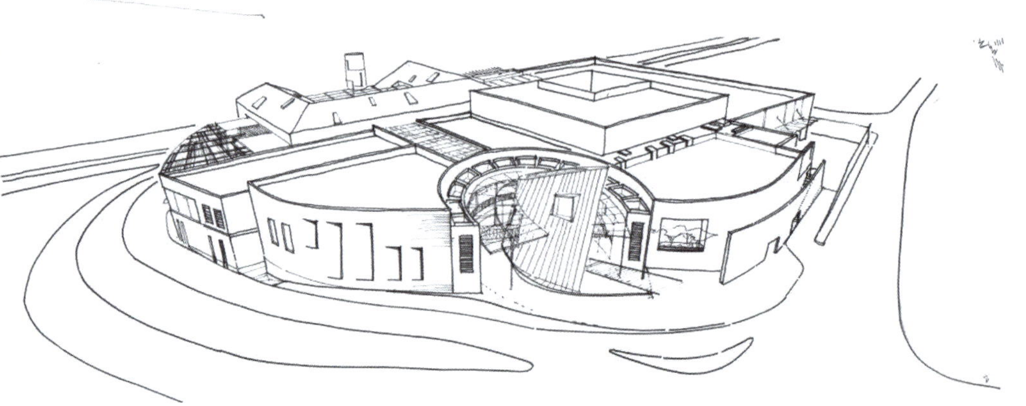

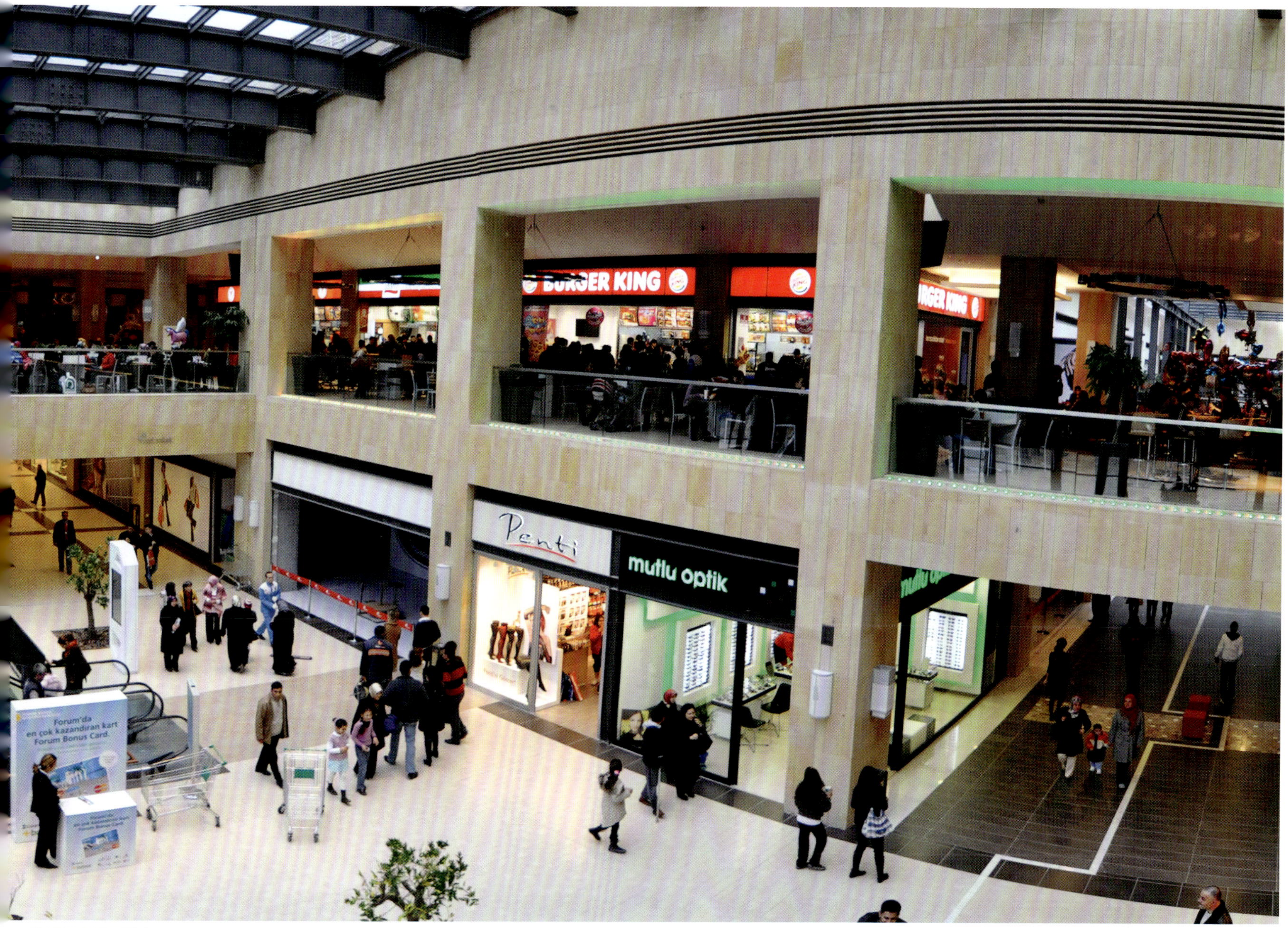
BURGER KING
Penti
mutlu optik
Forum'da
en çok kazandıran kart
Forum Bonus Card.

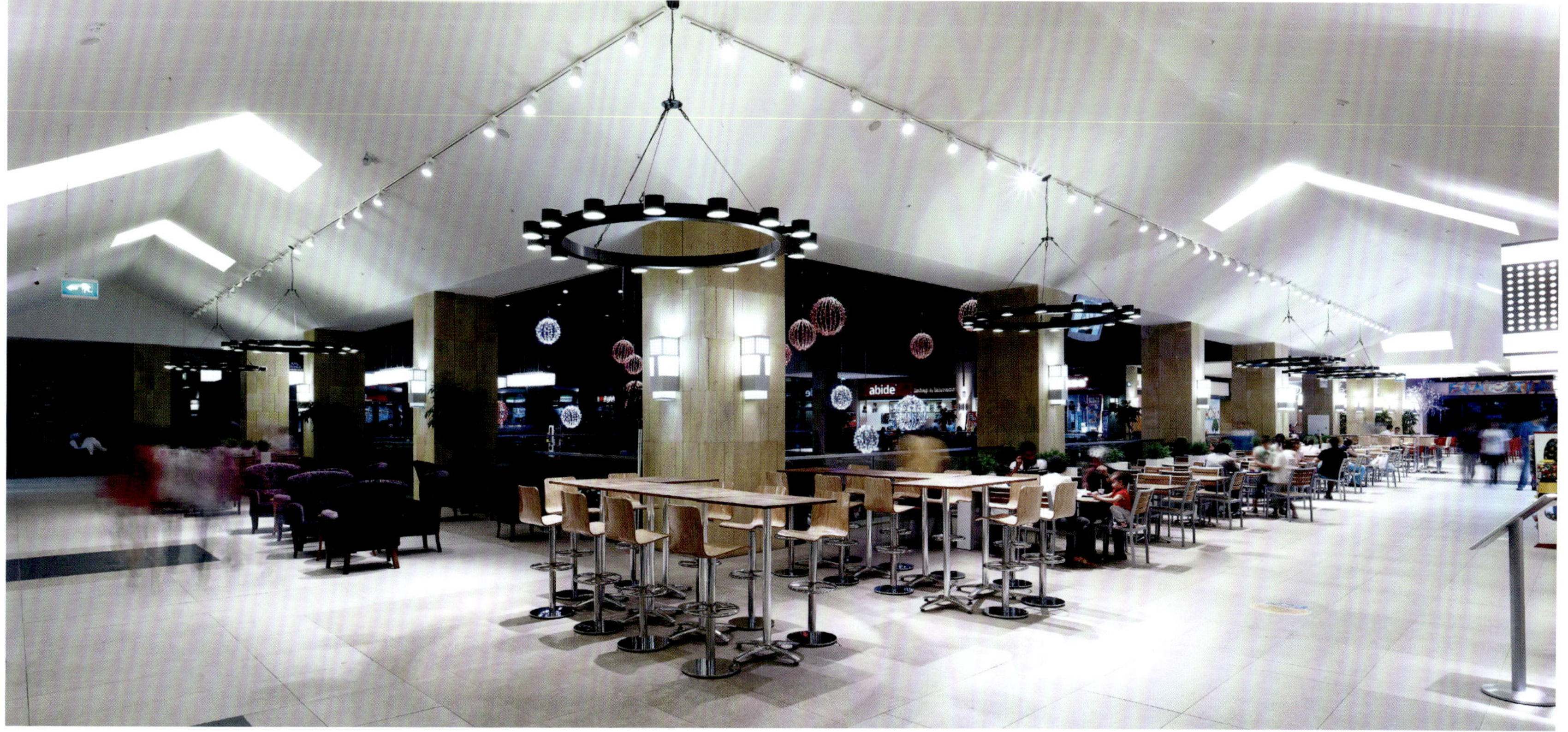

abide

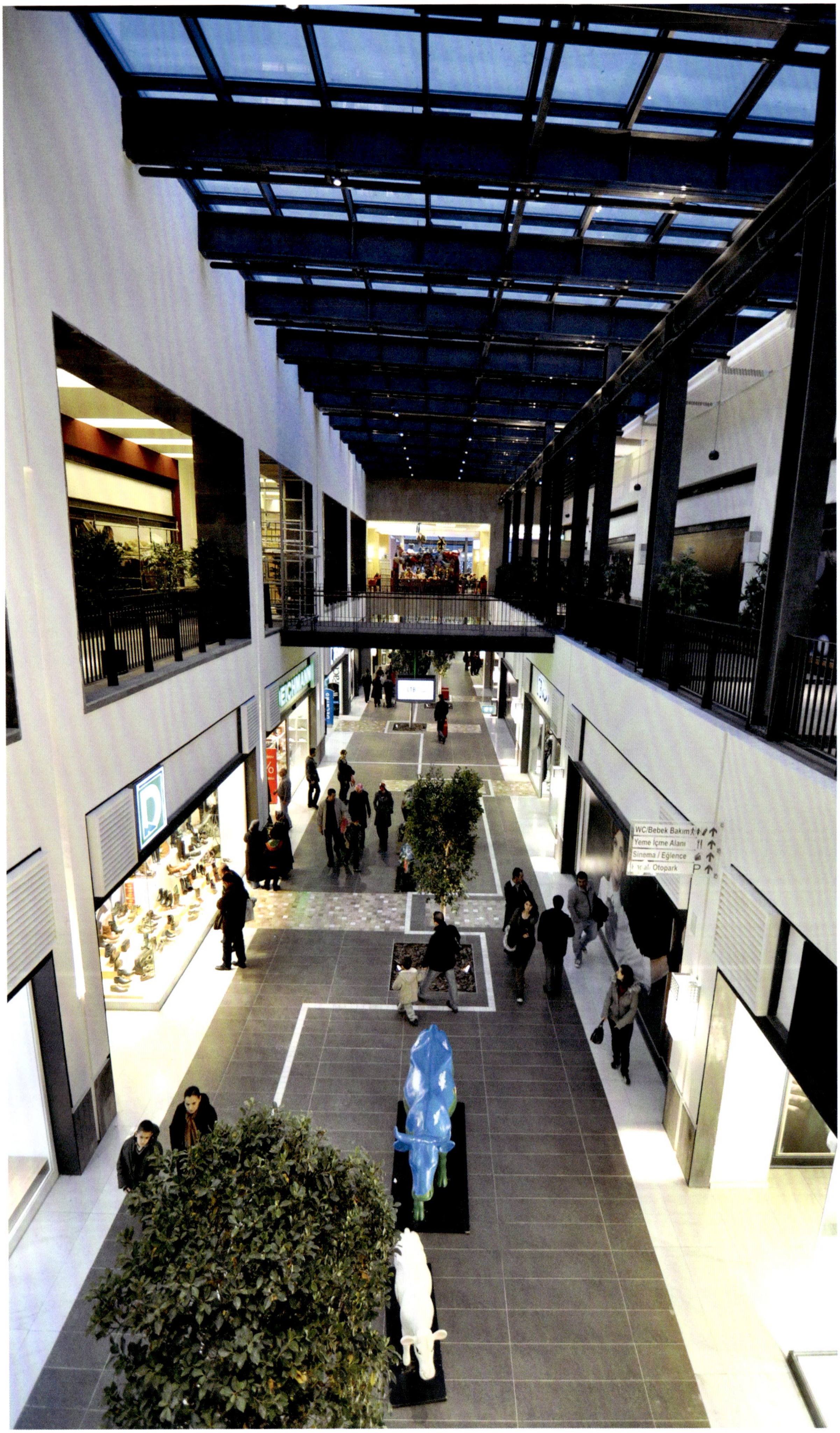
WC/Bebek Bakım
Yeme İçme Alanı
Sinema / Eğlence
Otopark
P

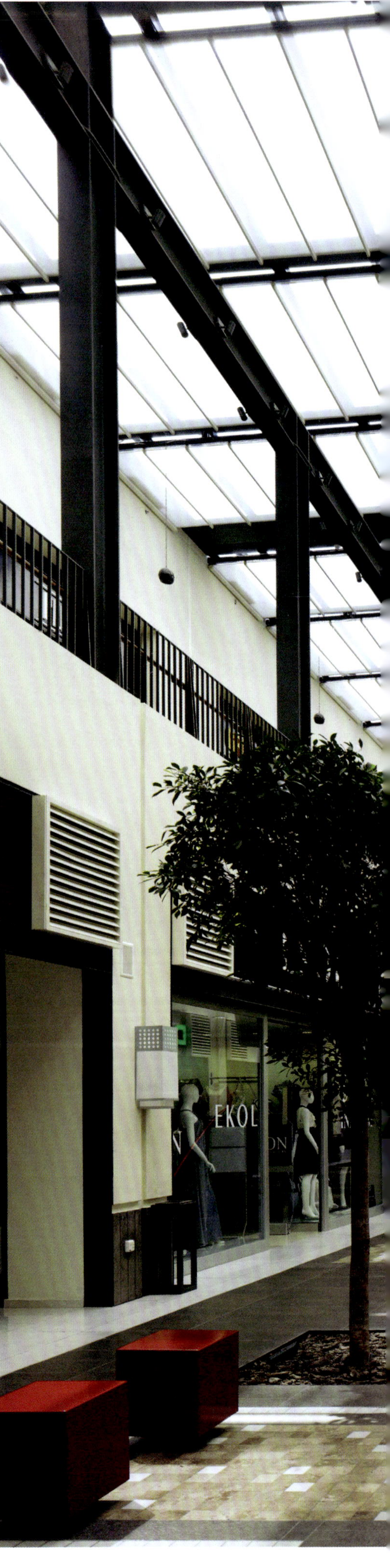

EKOL

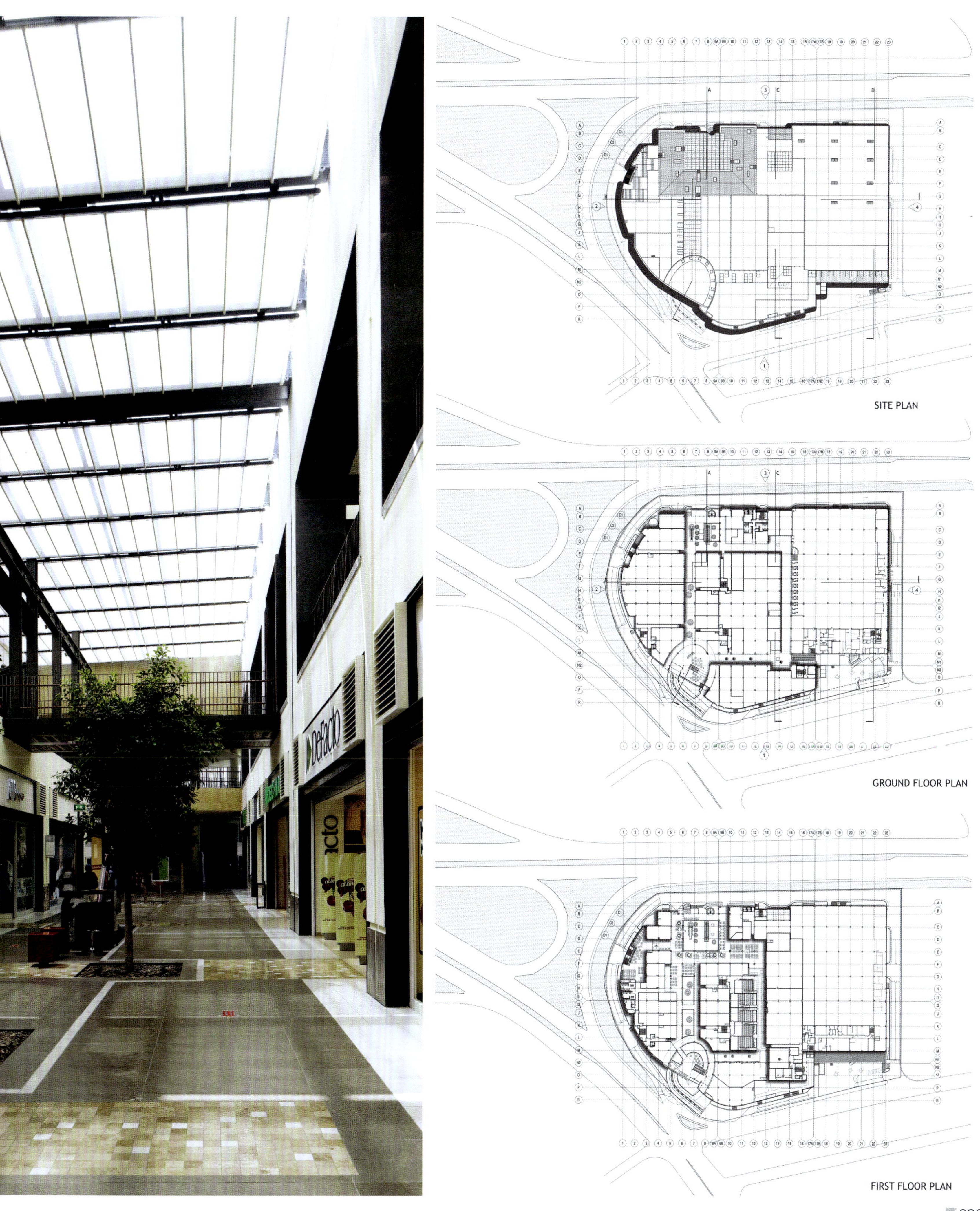
DeFacto
SITE PLAN
GROUND FLOOR PLAN
FIRST FLOOR PLAN

kipa HİPERMARKET
WC/Bebek Bakım
ATM / İlkyardım
Yönetim Ofisleri
hatemoğlu
1924
@html

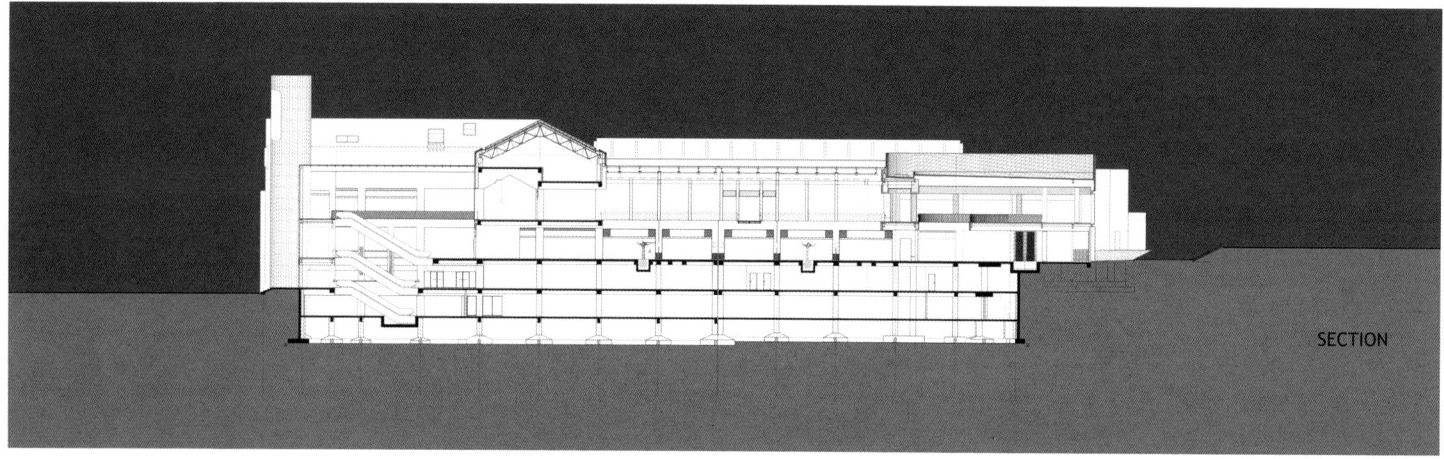
SECTION

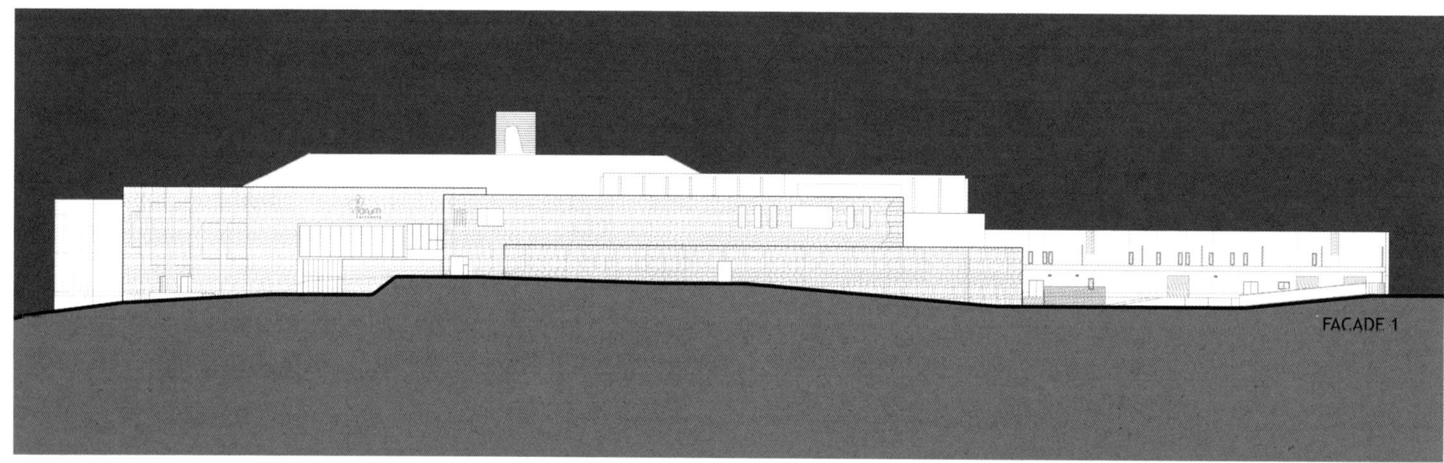
FACADE 1

FACADE 2

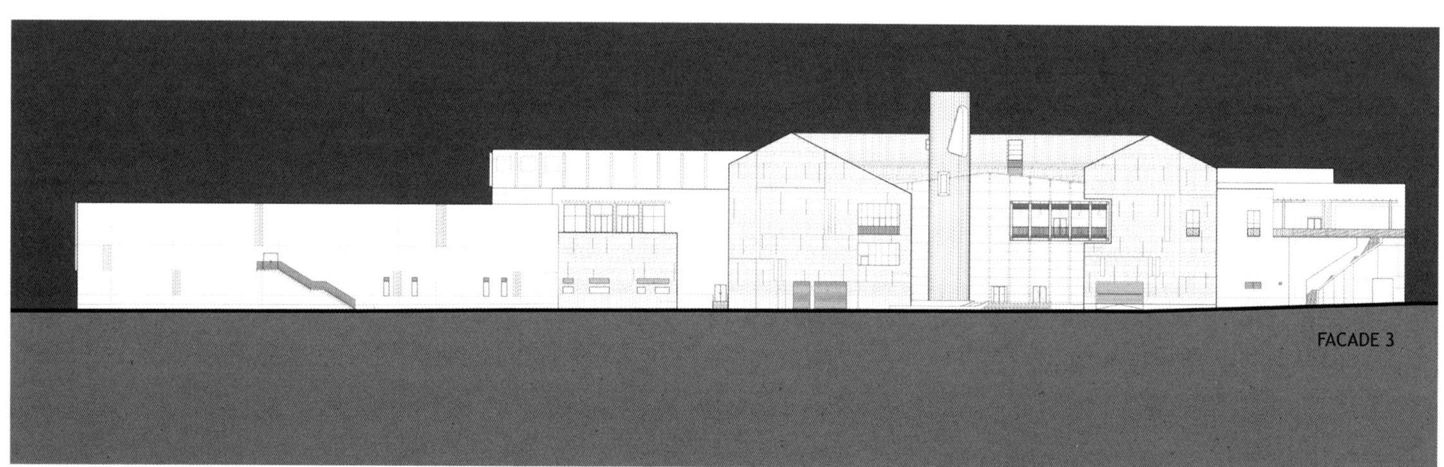
FACADE 3

# The New Gerngross
# 新葛罗斯商场

The atrium, which unites the horizontal and vertical visual axes, has becomes the store's new nucleus and central orientation point. The ceilings of the individual levels are rounded and "rolled" upwards. In this way, the atrium expands up towards the retail levels, appearing larger and more open than the atrium in its previous incarnation. The atrium and retail areas seem to melt together, bringing the individual levels into a mutual dialogue and making the store into a special continuum.

The retail areas of the new store are distributed like ice floes on each story. These floes are marked by a change in floor covering and ceiling material, thereby forming a star - shaped pattern of paths on both the ceiling and the floor.

In this way, the materials on the floor and ceiling facilitate orientation within the building. Paths and retail areas are explicitly marked and visible from afar:

The facade of the new building forms the logical continuation of the interior: the ice-floe theme of the interior space continues on the new facade. To this end, large-scale, amorphous colour fields were applied. Leaving some space in between, an ornamentally designed, semi-transparent white area was then attached. Together, these layers form a conglomerate of light and colour.

**Location**

Mariahilferstraße 42-48, Vienna, Austria

**Area**

52,500 sq.m.

**Main materials**

Wall gypsum, flooring tiles, ceiling perforated metal areas, ornamental patterns

**Company**

LOVE architecture

**Designer**

LOVE architecture

**Photographer**

Jasmin Schuller, Bruno Klomfar

中庭结合了横向和纵向的视觉轴，成为百货店新的核心和中心定位点。个别层面的天花板是圆形的而且向上“轧”。通过这种方式，中庭向上往零售层面扩张，这样便出现了比以前更大、更加开放的中庭。中庭和零售区域似乎融在了一起，把个别层面带进了一个相互对话中，并使百货店成为一个特殊的有机连续体。

新百货店的零售区域的分布像飘在每层楼上的浮冰。这些浮冰的标志是楼面覆面层和天花板材料的变化，从而在天花板和地板上形成了星形格局的路径。因此，地板和天花板的材料有利于指引建筑物内的方向，明确标记了路径和零售领域，而且从远处可见。

新建筑的立面形成内部的合乎逻辑的延续：室内空间的浮冰主题延续在新的立面上。为此，大规模的，无定形的颜色领域被应用。在之间留一些空间，附加一个用作装饰设计的半透明的白色区域。总之，这些层形成一个光线和色彩的混合物。

SATURN
Gerngross
SATURN
SPORTS EXPERTS
MERKUR
Akakiko
P
PALMERS
promod

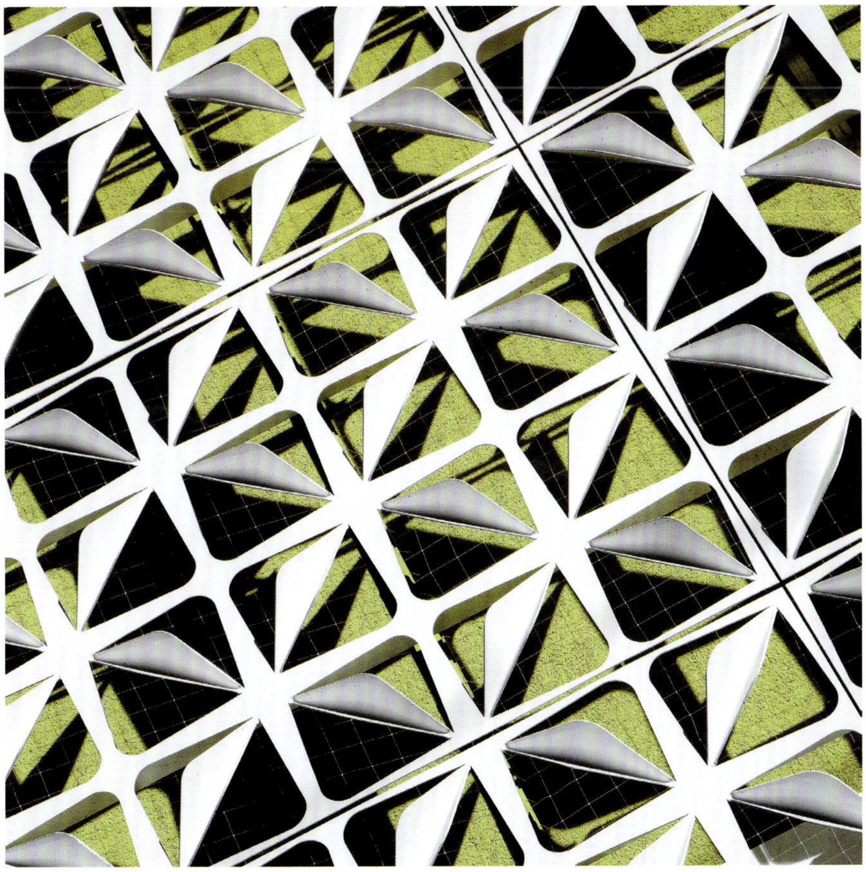

ERÖFFNUNG 21. – 23. OKT.
Gerngross
ZARA
ZARA

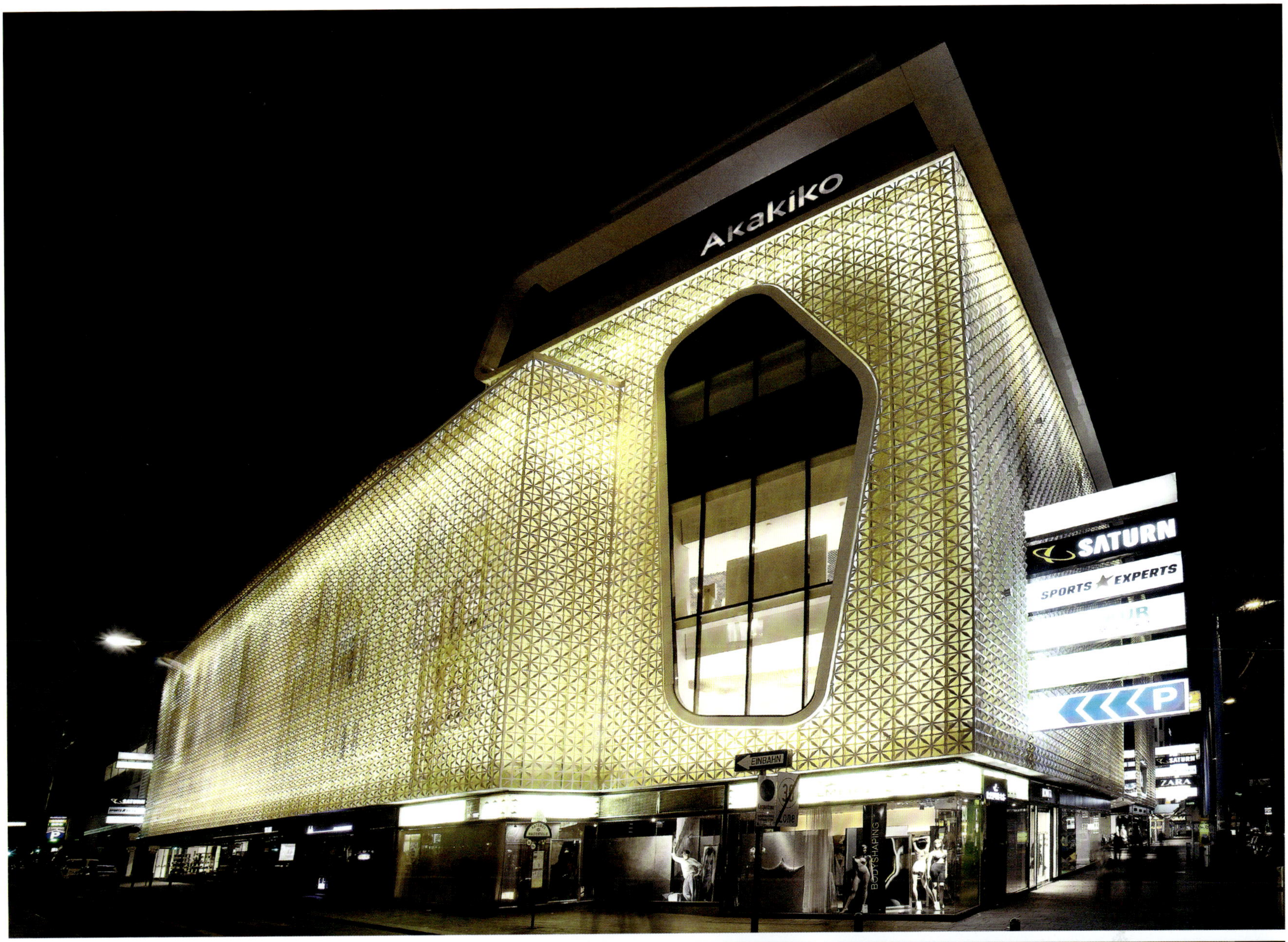
Akakiko
SATURN
SPORTS EXPERTS
EINBAHN
BODYSHAPING

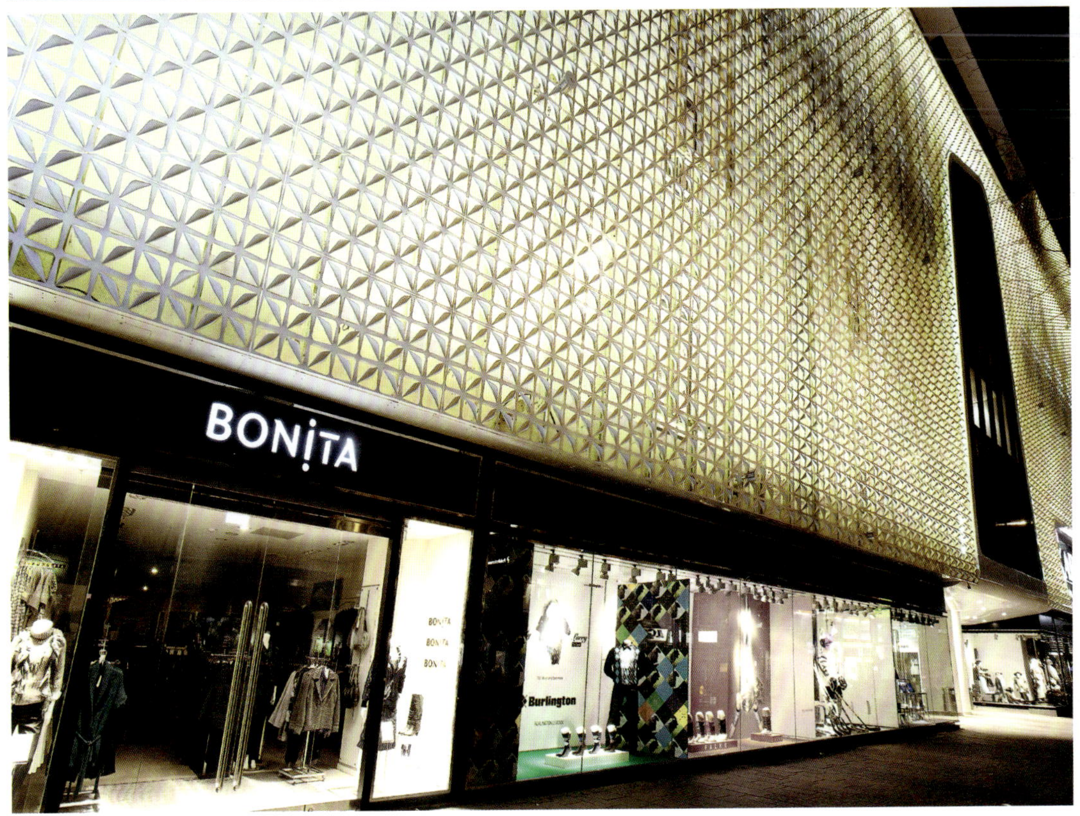

BONITA
Burlington

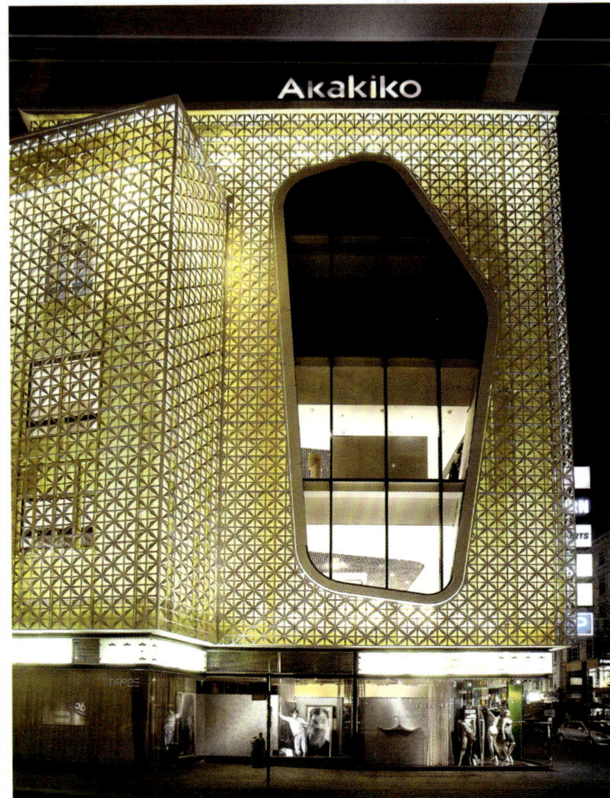

Akakiko

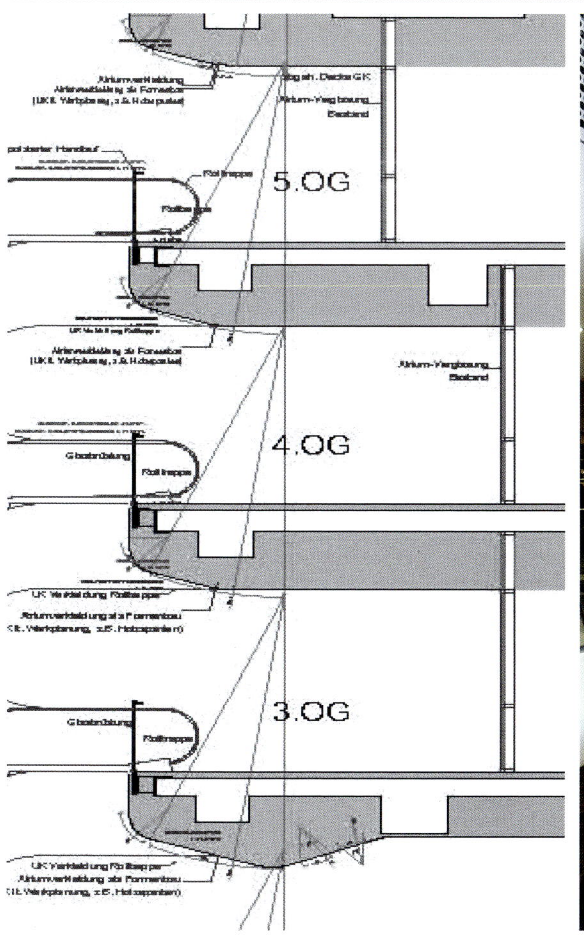
5.OG
4.OG
3.OG

# Chadstone West Mall

## 查斯特西商场

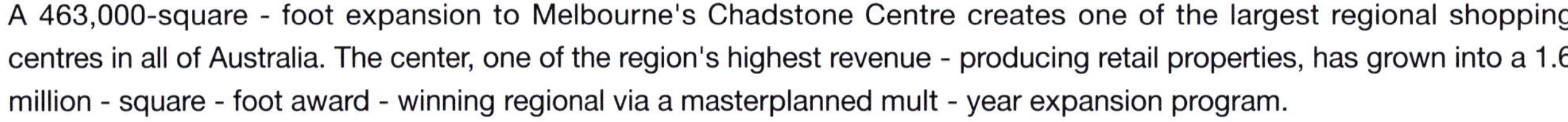

A 463,000-square - foot expansion to Melbourne's Chadstone Centre creates one of the largest regional shopping centres in all of Australia. The center, one of the region's highest revenue - producing retail properties, has grown into a 1.6 million - square - foot award - winning regional via a masterplanned mult - year expansion program.

The initial phase adds a new two - level high fashion district at the southern end of the centre. Refined landscaping and bronze/stone detailing provides a new landmark identity and front door. Inside, the space is defined by a distinctive spiral of glass that culminates in seashell - shaped skylights.

The second phase adds a new entertainment zone at the centre's northern end. A new entry affords direct access to cinemas, restaurants and other entertainment venues. The existing eight-screen cinema is revamped and expanded to 19 screens on two levels. A lower - level cinema is reconfigured and expanded to 13 screens. Two new parking structures complete the expansion, increasing parking capacity from 5,600 to 9,500 cars.

**Location**
Melbourne, Australia

**Area**
43,012.7 sq.m.

**Company**
RTKL Associates Inc.

**Designer**
RTKL Associates Inc.

**Photographer**
Aaron Pocock

墨尔本查斯特商场中心扩展了463,000平方英尺创建了澳大利亚所有最大的区域购物中心之一。这个中心，是该地区收入最高的零售物业之一，已经成长为一个160万平方英尺的屡获殊荣的区域建筑，是通过规划多年的扩张项目。

初始阶段在该中心的南端建立一个新的两层楼的高级时装区。精制的美化和青铜或石头雕刻，提供了一个新的具有里程碑意义的标志和前门。内部空间被一个独特的玻璃螺旋所定位，玻璃螺旋的顶端是一个贝壳形的天窗。

第二阶段在该中心的北端增加了一个新的娱乐区。一个新的入口可以直接进入电影院，餐馆和其他娱乐场所。原有8个屏幕的电影院已改革和扩大到19个屏幕，分布在两层楼上。较低层的电影院重新配置，并扩大至13个屏幕。完成扩展了两个新的停车场结构，使停车容量从5,600辆车增加到9,500辆车。

Capital
Kitchen.

TIFFANY & CO.
TIFFANY & CO
Capital Kitchen

BURBERRY
TIFFANY & Co.
RALPH LAUREN
TIFFANY & Co.

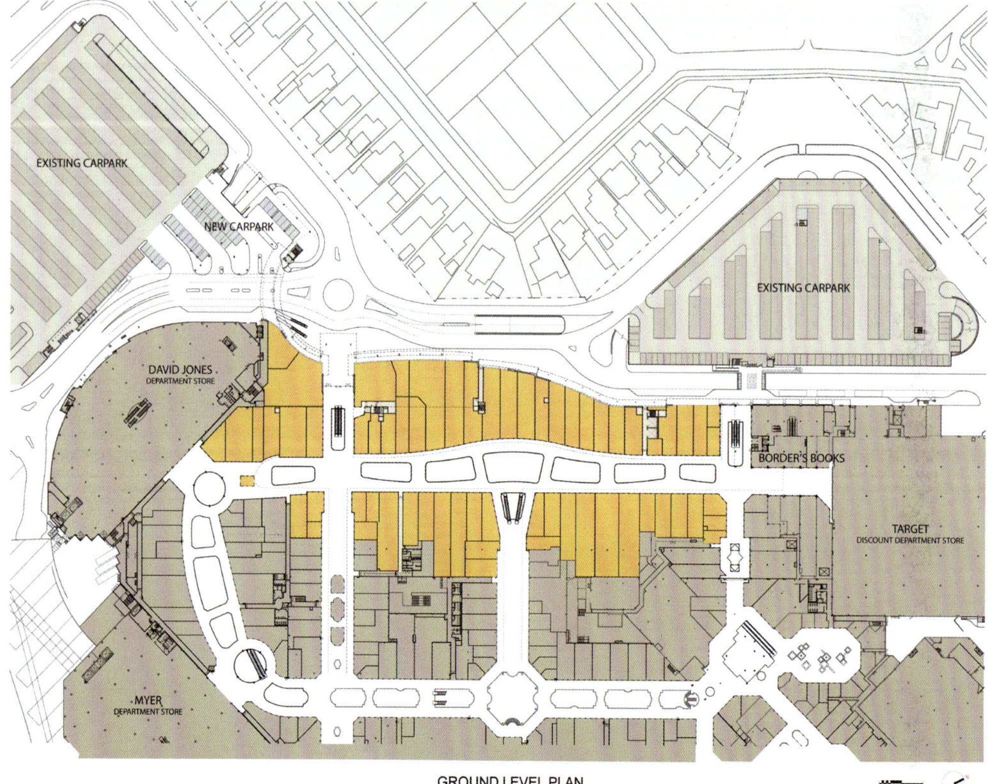


GROUND LEVEL PLAN

ESPRIT
collection
ESPRIT

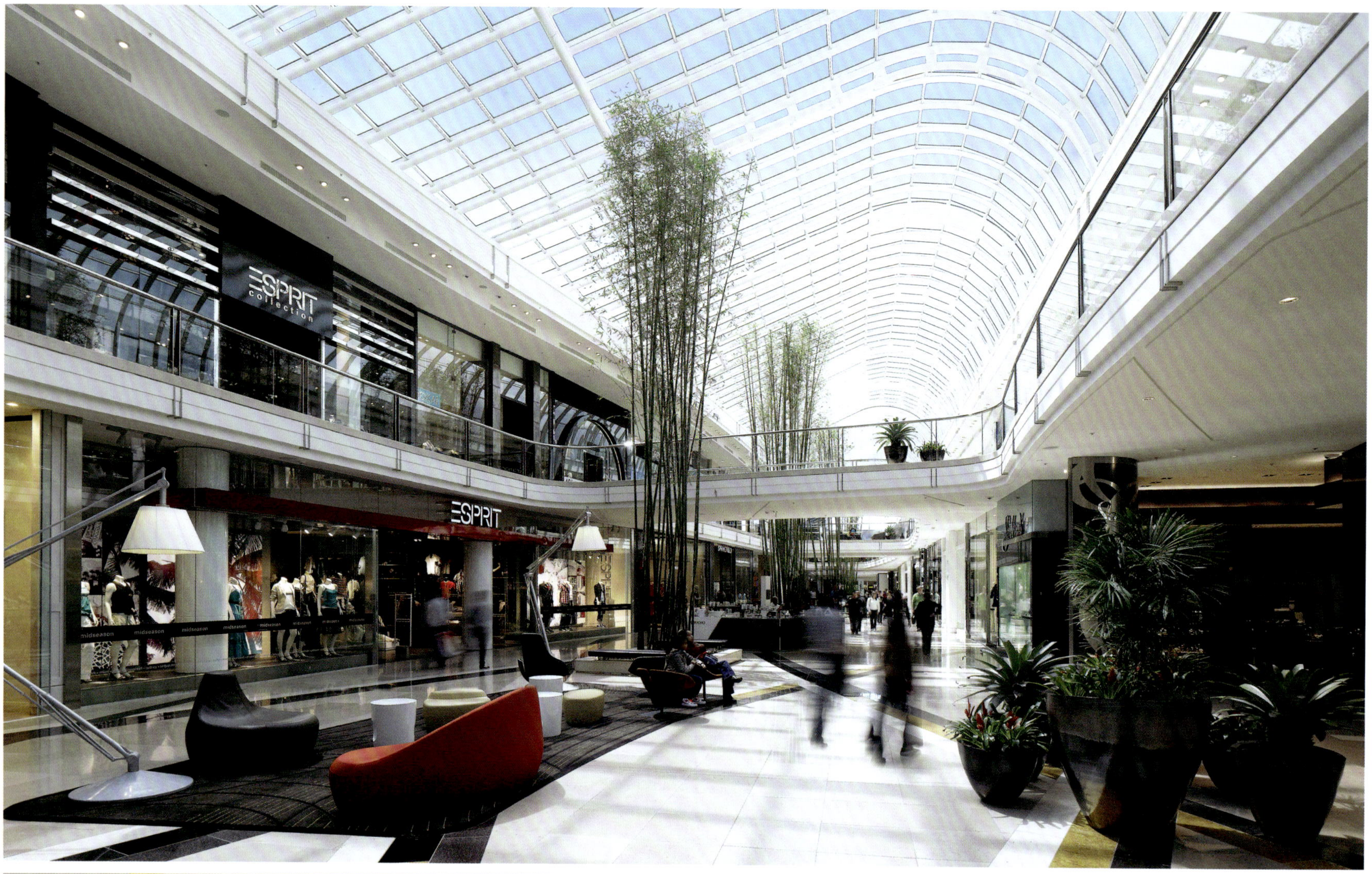
ESPRIT
collection
ESPRIT

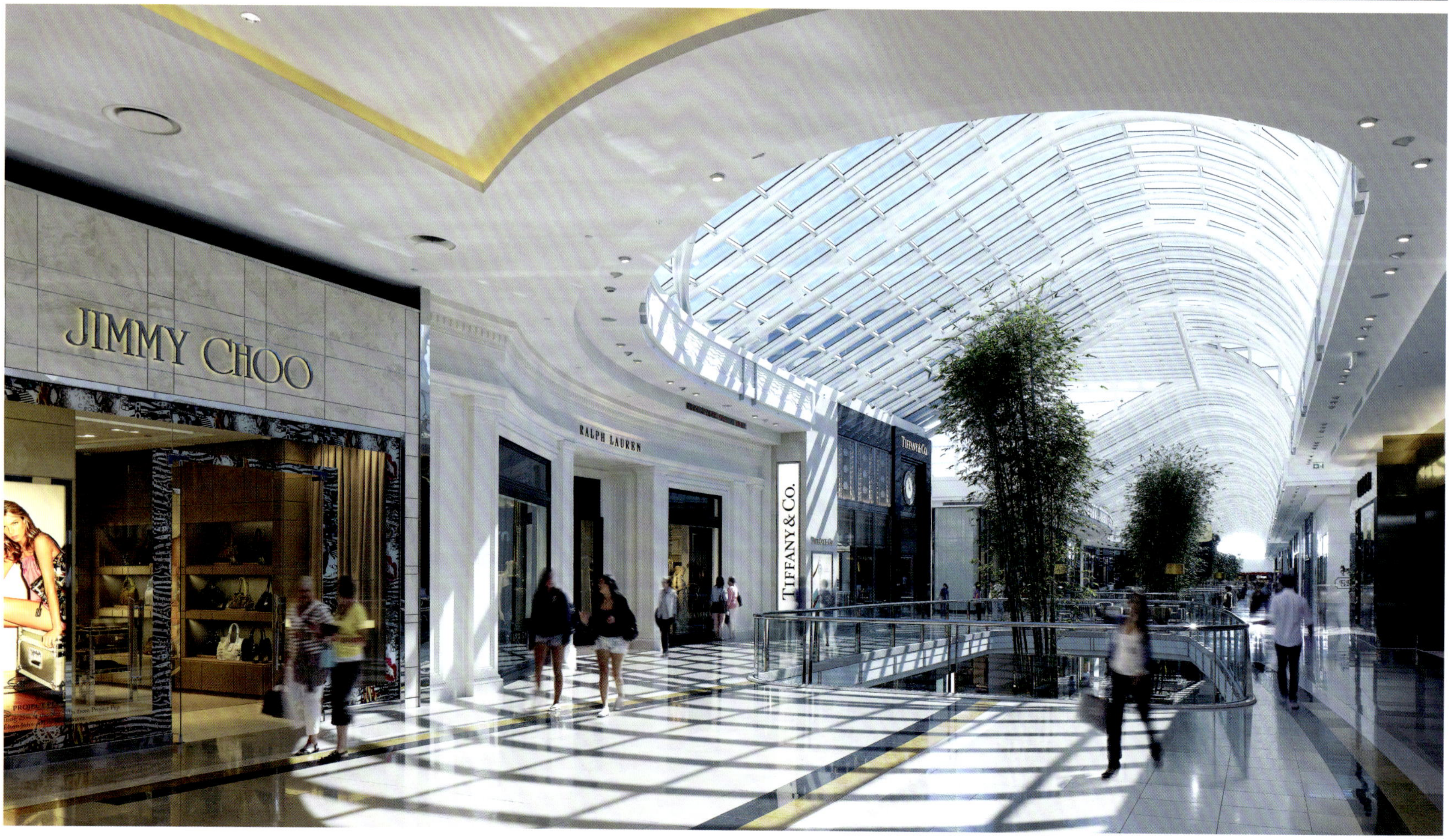
JIMMY CHOO
RALPH LAUREN
TIFFANY&Co.

BORDERS
ANTON

# Crystals at CityCenter

## 市中心Crystals购物中心

Rockwell Group envisioned Crystals as an abstracted 21st century park to fit into this new larger-than-life urban center in Las Vegas. The organic, curvilinear vocabulary of the interior architecture reinvents and re-imagines the idea of the central urban park as a social gathering place for shopping and dining.

Interpretations and abstractions of nature animate the scenery that complements the sharp angles of Studio Daniel Libeskind's crystalline exterior, with a glowing three-story wooden sculpture inspired by a modern tree house. Floor-to-ceiling columns and trellises above are abundant with hanging plants as a reinterpretation of a gazebo and a dynamic flower carpet at the node of the three promenades in Crystals that transforms with the seasons. Natural and tactile materials abound in response to CityCenter's overall LEED Gold aspirations, including the flowing 24-foot grand bamboo stair, inspired by Rome's Spanish steps.

**Location**
Las Vegas, Nevada

**Area**
46,451 sq.m.

**Main materials**
Bamboo

**company**
Rockwell Group

**Designer**
David Rockwell, David Mexico

**Photographer**
Scott Frances

罗克韦尔集团设想此项目Crystals是一个抽象的21世纪公园，来适应这个新的有传奇色彩的拉斯维加斯的城市中心。室内建筑的有机的曲线词汇改造并重新想象城市中央公园的理念成为购物和餐饮的一个社交聚集地。

自然的诠释和抽象性使补充了李伯斯金工作室透明外观的尖角的风景生动起来，加上一个光辉的3层楼高的木制雕塑，这个灵感来自一个现代化的树屋。上面有大量的落地柱和棚架，带有悬挂植物，作为对一个露台和一个动感的花地毯的重新解释，这花地毯位于本项目的3条长廊的交叉处，并随着不同季节而改变。天然和有形的材料比比皆是，来回应CityCenter的整体LEED金牌的抱负，包括24英尺的豪华平滑竹楼梯，此灵感来自罗马的西班牙台阶。

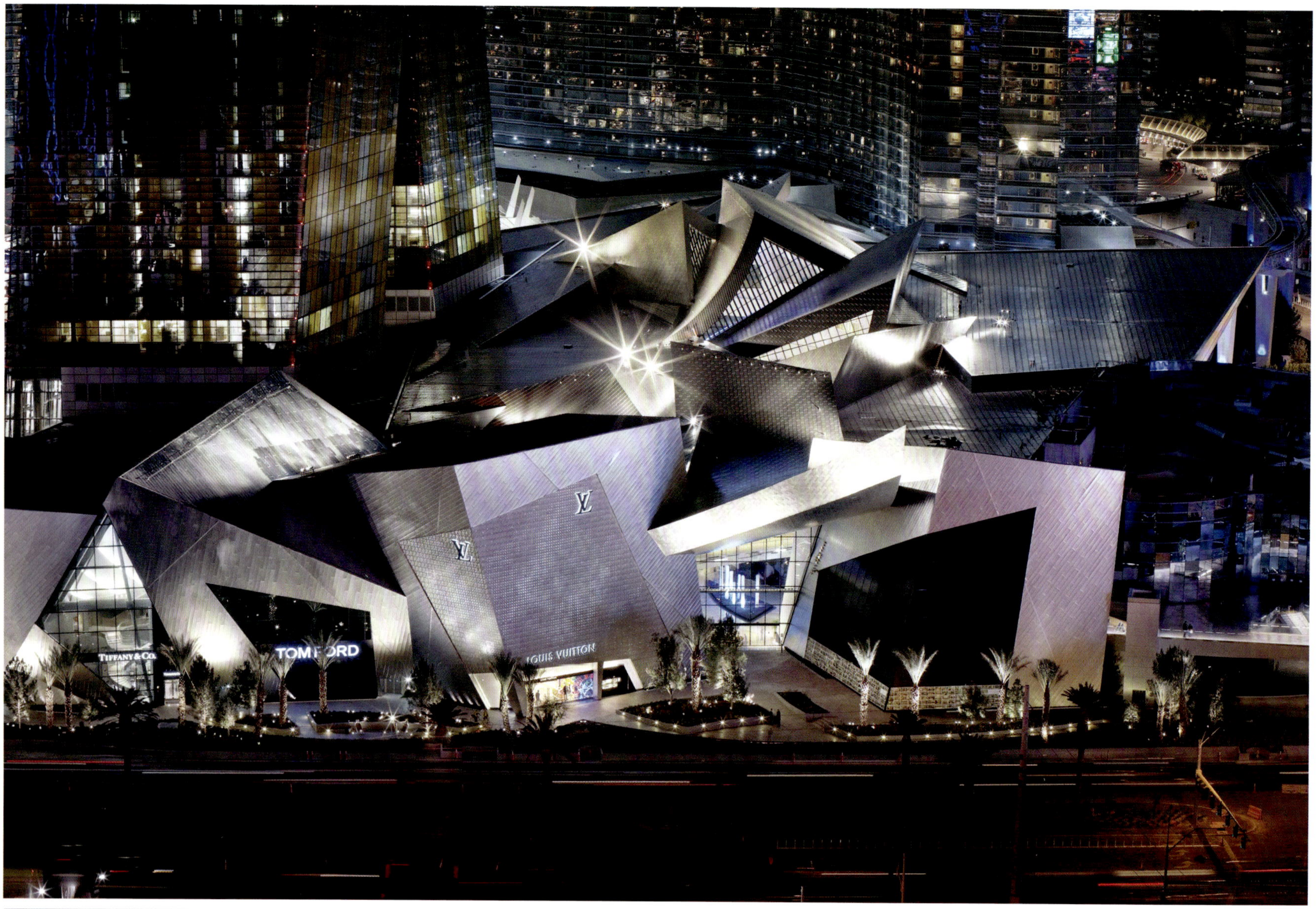
TIFFANY&Co.
TOM FORD
LOUIS VUITTON

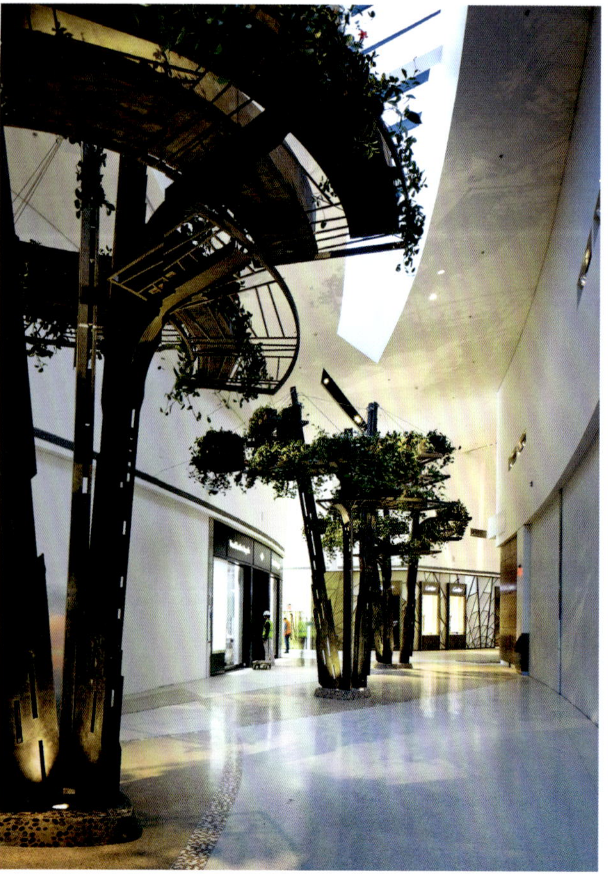

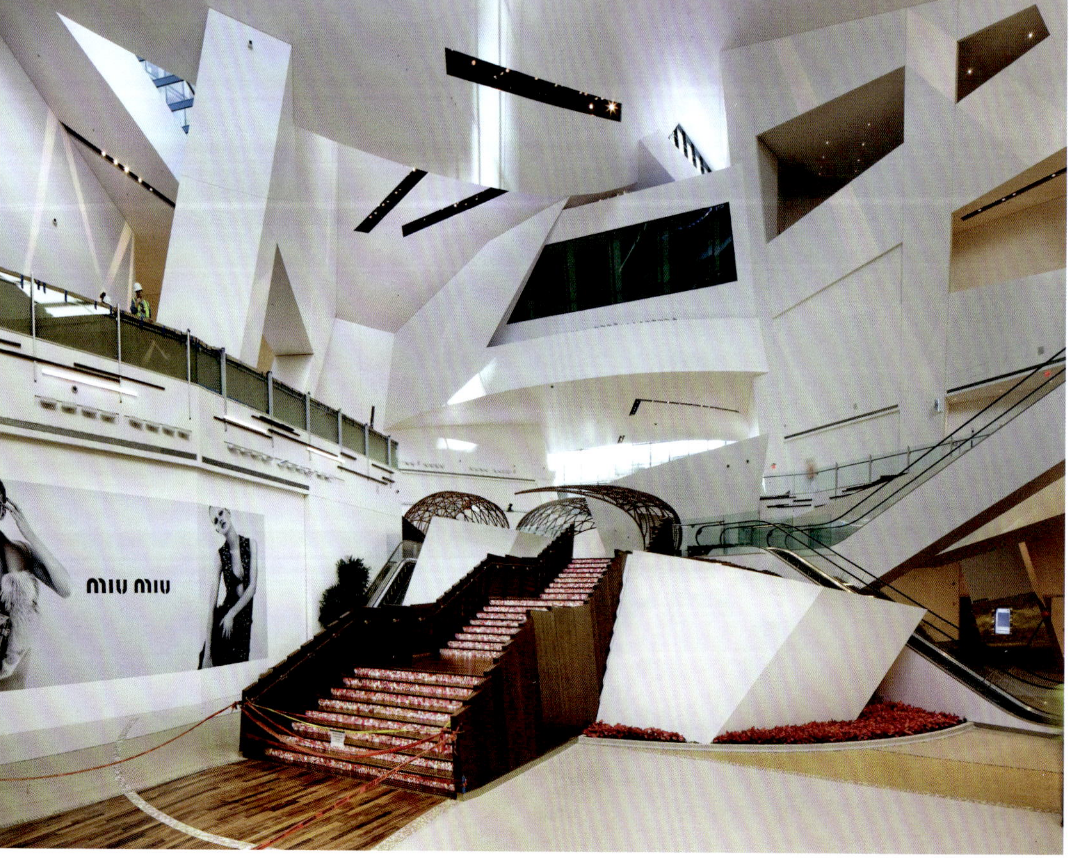

miu miu

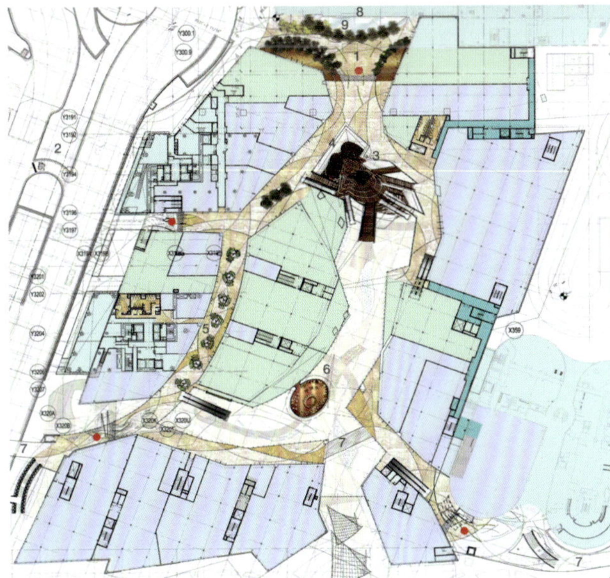

1 Casino Square
2 Casino Boulevard
3 Grand Stair
4 Pods
5 Hanging Gardens
6 Tree House
7 Park Bridges (3)
8 Casino Lobby
9 Reflective Pool
Retail
Entertainment, Food & Beverage
Main Entrances (4)

1 Crystals Valet
2 Oyster
3 Grand Stair
4 Water Vortex
5 Flower Carpet
6 Tree House Base - Concierge
7 Ice Totems
8 Tree Root
Retail
Entertainment, Food & Beverage
Main Entrances (3)

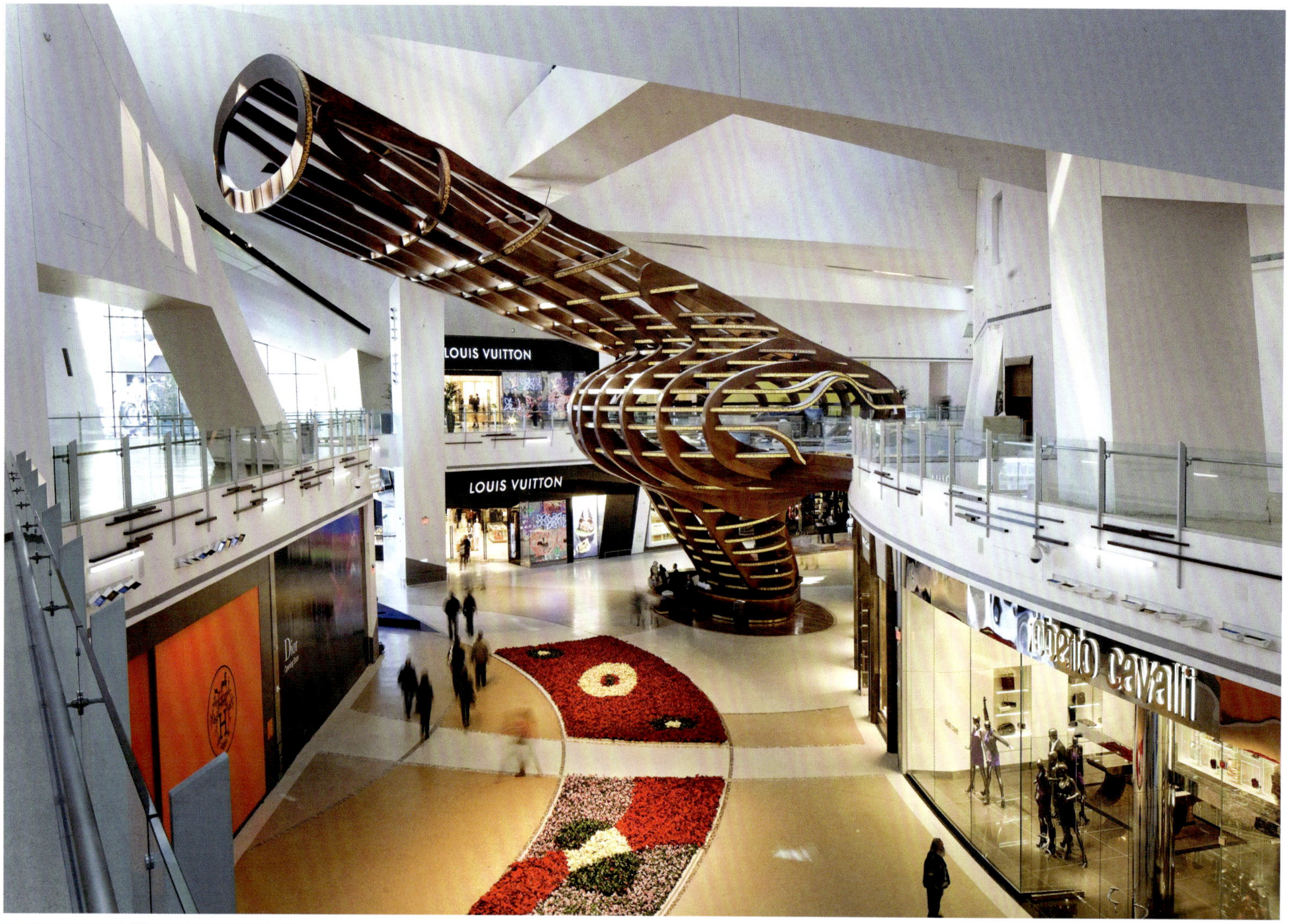
LOUIS VUITTON
LOUIS VUITTON
Dior

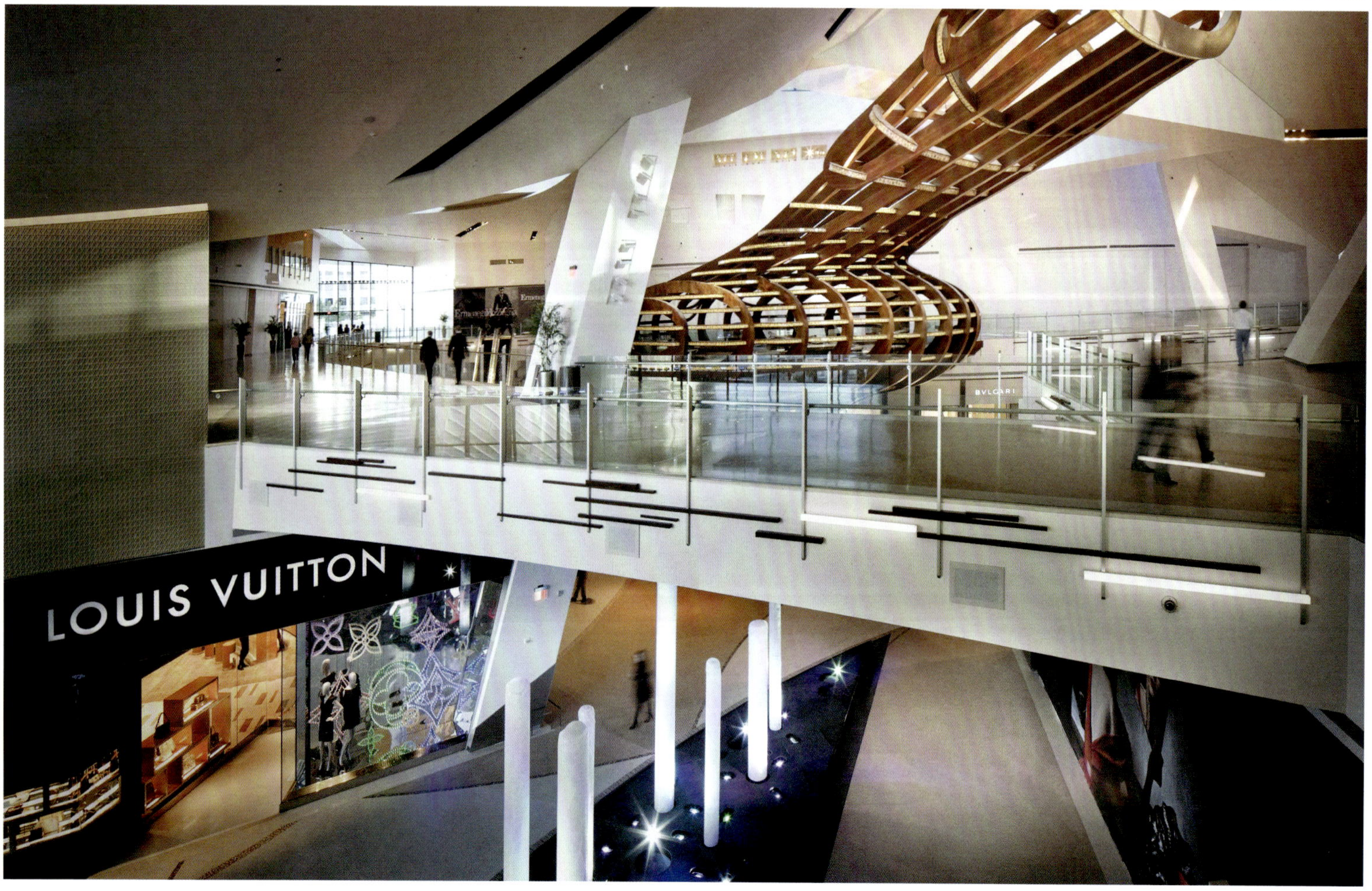
LOUIS VUITTON

LOUIS VUITTON
LOUIS VUITTON

# Vershina Trade & Entertainment Centre Surgut

# 苏尔古特Vershina贸易和娱乐中心

The "Vershina" Trade and Entertainment Centre is the first five - star international shopping centre in Surgut and offers space for retail, extreme sports areas, dance studios, restaurants, bars and an underground night - club. On a total gross floor area of 35,000 square meters spread over eight storeys, the building provides around the clock activities for all visitors, young and old. The concept for the building is based on a dialectical play between dark and light, solid and transparent, open and closed. The building is equipped with an extensive exterior lighting scheme and its glass facade forms a screen onto which moving advertisements are projected. The primary cuts in the facade divide this basic mass into "sharp volumes" that allow daylight in, and radiate artificial light out at night. These cuts are accompanied by secondary cuts in the facades, the so-called "lines of light". The primary and secondary cuts in the building skin will allow for the building to become a beacon of light during the nine dark winter months in Surgut.

**Location**
Surgut, Tumen Region, Russian Federation

**Area**
37,050 sq.m.

**Main materials**
Concrete, natural stone, glass, synthetic, etc.

**company**
Erick van Egeraat

**Designer**
Erick van Egeraat

**Photographer**
Alexey Naroditskiy

**Company**
wwwdvip-hk co.,ltd

“Vershina”贸易和娱乐中心是苏尔古特第一个五星级国际购物中心，拥有零售区、极限运动区、舞蹈室、餐厅、酒吧和地下夜总会。这栋建筑总面积为35000平方米，涵盖八层，为所有游客，无论男女老少，提供二十四小时的活动。建筑的概念是基于明与暗、实心与透明、开放与封闭之间的辩证发挥。建筑配备有一个外部照明体系，它的玻璃门面形成一面屏幕，晚上移动广告可投影在上面。门面上的主要切口把这个基本体块划分为多个“轮廓分明的单体”，这样白天时自然光可以照入室内，而到了晚上，可以往外映射出人造光。这些切口在门面上附有次要切口，即是所谓的“光线条”。建筑表面上主要的和次要的切口使建筑在苏尔古特长达九个月的暗冬里成为一座灯塔。

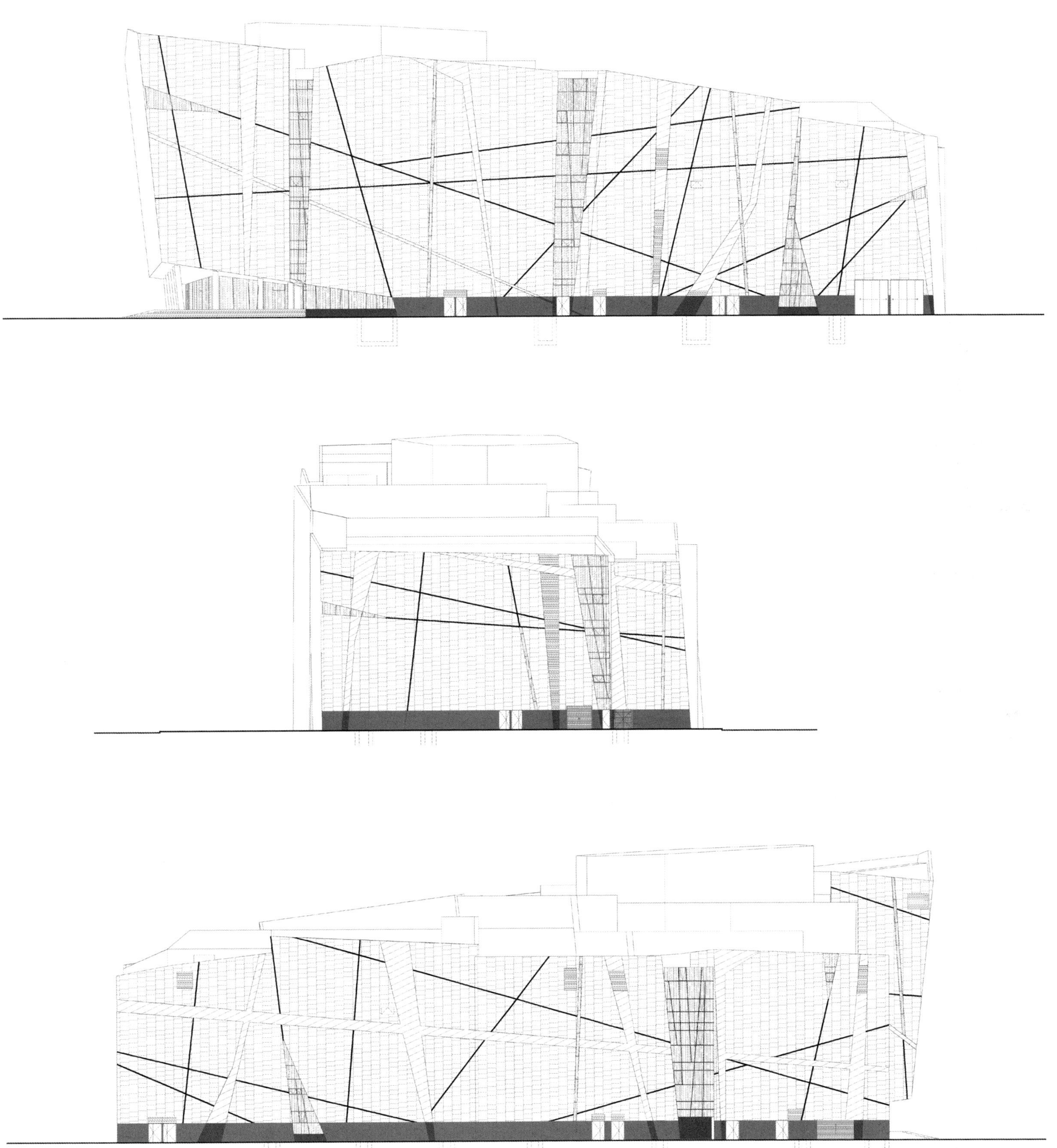

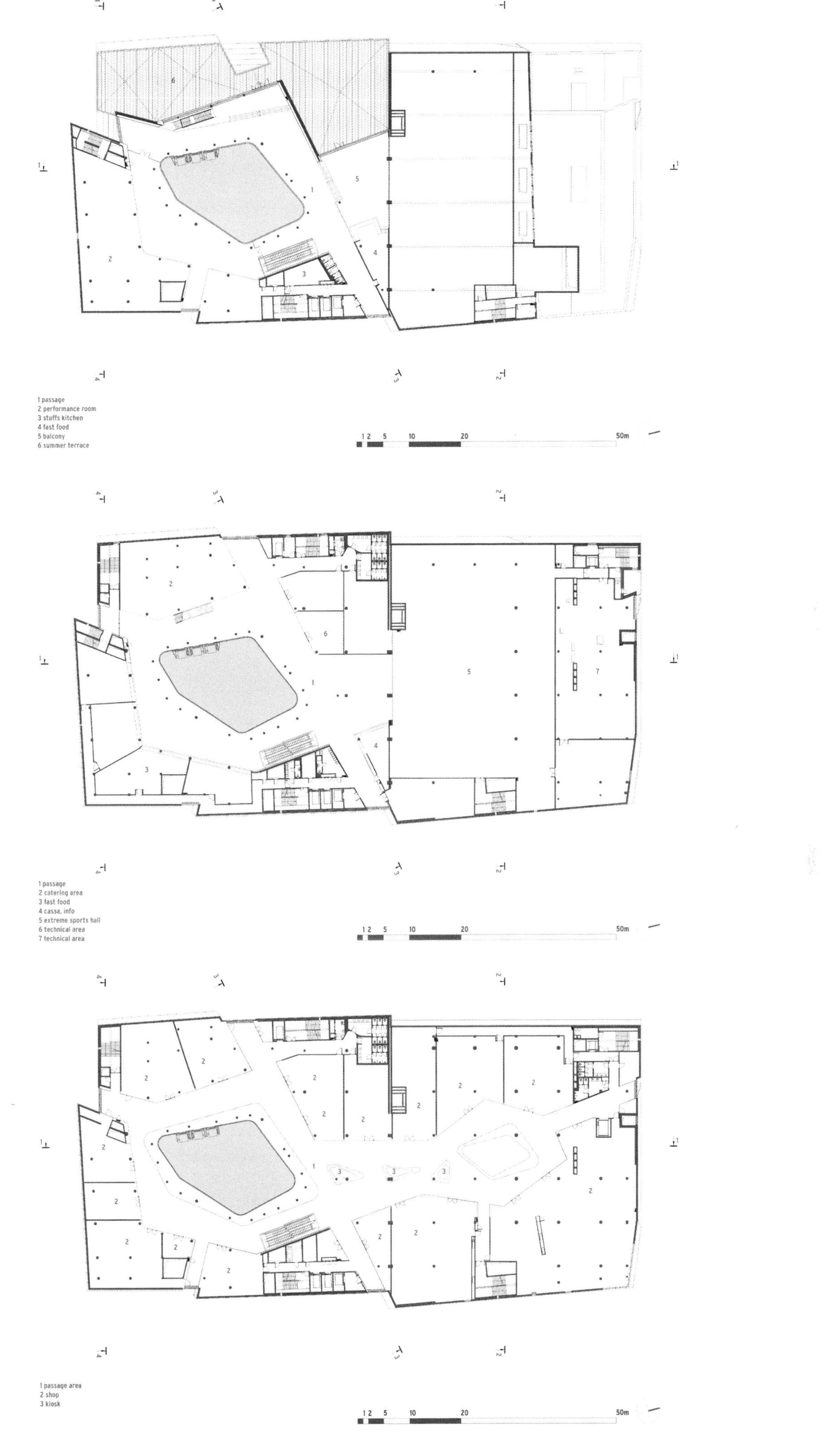
1 passage
2 performance room
3 stuffs kitchen
4 fast food
5 balcony
6 summer terrace
1 2 5 10 20 50m
1 passage
2 catering area
3 fast food
4 cassa, info
5 extreme sports hall
6 technical area
7 technical area
1 2 5 10 20 50m
1 passage area
2 shop
3 kiosk
1 2 5 10 20 50m

ОТК
-РЫ-
0 ЭТАЖ
1 ЭТАЖ
2 ЭТАЖ
3 ЭТАЖ
O'STIN
FASHION
LACOSTE
TOP SECRET
CONCEPT CLUB
OGGI
adidas
YVES ROCHER
BOUTIQUE SWAROVSKI
LOVE REPUBLIC
GLENFIELD
4 ЭТАЖ
CHESTER
Серебряный ХИТ
Gloria Jeans
ACOOLA
lady & gentleman CITY
SOLENNE COLLEZIONE
Colours&Style
КОМФОРТ
5 ЭТАЖ
Кинотеатр
SNACK BAR
Чао-Какао
ХОЛИДЕЙ ПИЦЦА
SUBWAY
JUICE Bar
Дольче Вита
6 ЭТАЖ
Семейный развлекательный центр
Летняя терраса
Burgerwood
5 ЭТАЖ
Kira Plastinina
7 ЭТАЖ
Летняя терраса

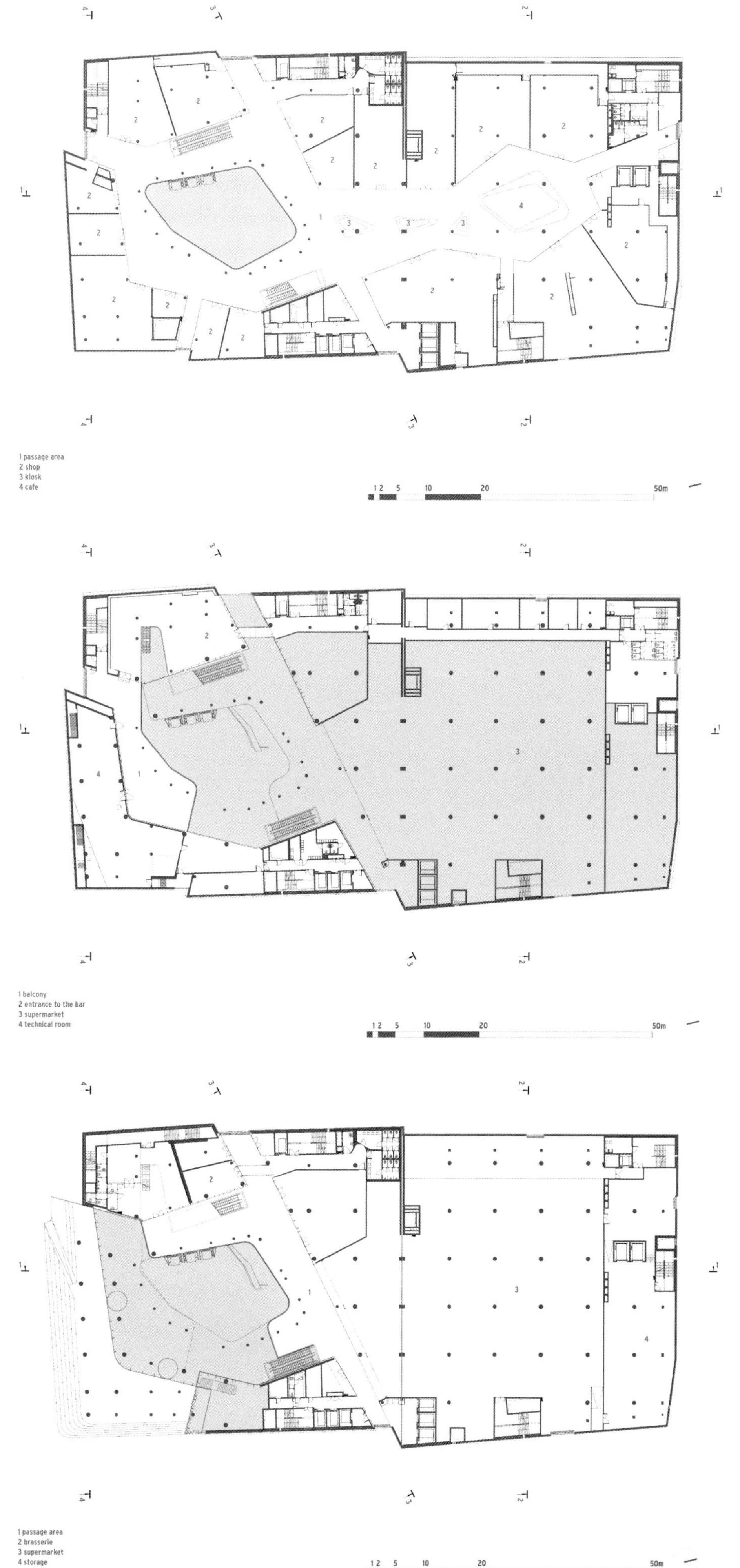
1 passage area
2 shop
3 kiosk
4 cafe
1 2 5 10 20 50m
1 balcony
2 entrance to the bar
3 supermarket
4 technical room
1 2 5 10 20 50m
1 passage area
2 brasserie
3 supermarket
4 storage
1 2 5 10 20 50m

ЛИФТ

1 entrance
2 entrance and atrium
3 entrance to the night club
4 ramp, car entrance
5 parking

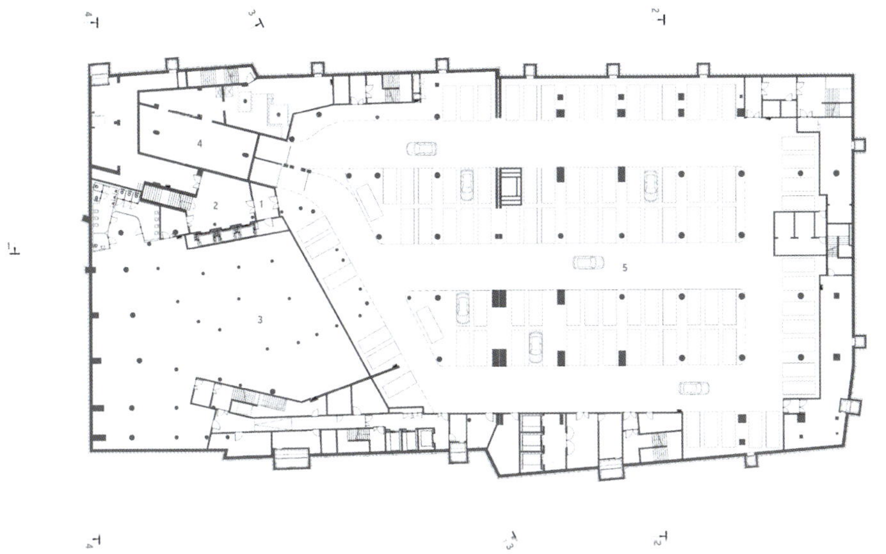

1 entrance
2 entrance and atrium
3 night club
4 ramp, car entrance
5 parking

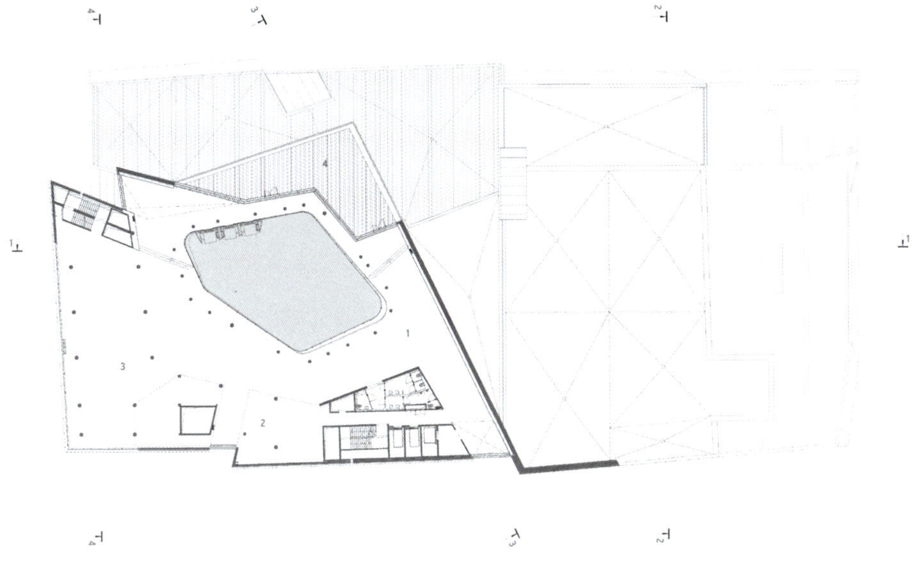

1 passage
2 kitchen
3 restaurant
4 summer terrace

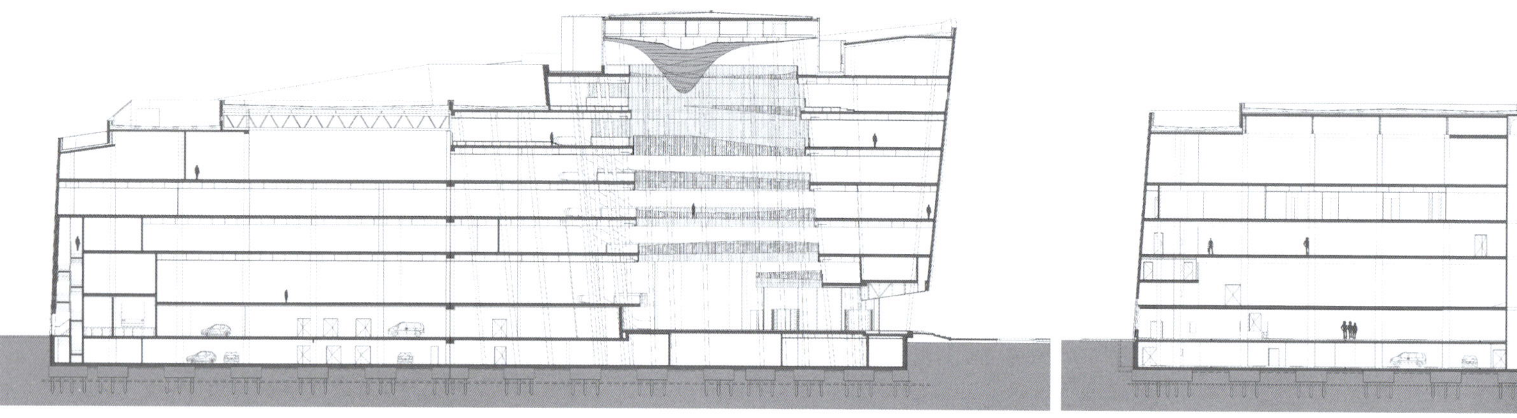

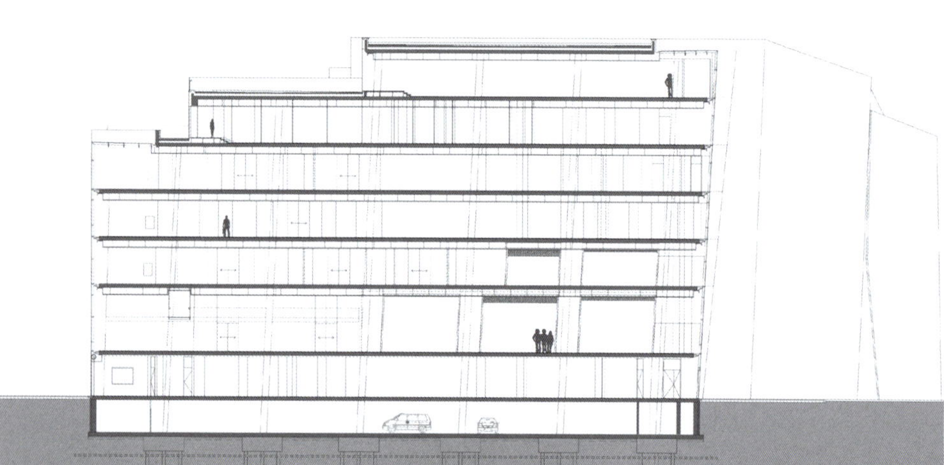

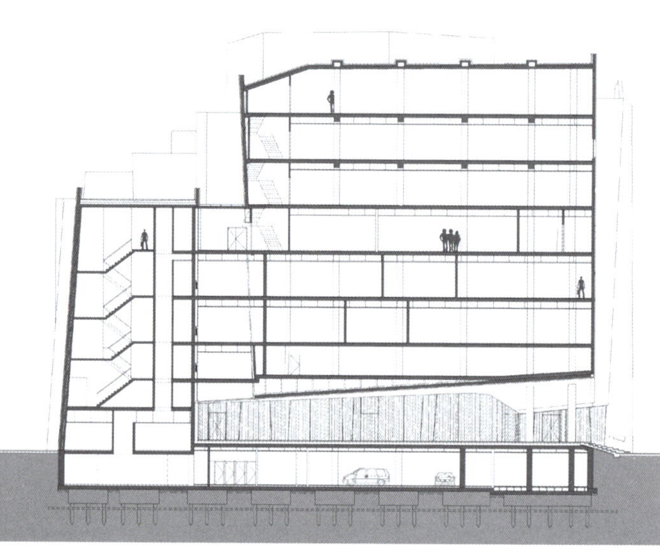

# Zlote Tarasy

## Zlote Tarasy购物中心

The roof is constructed of approximately 4,700 triangular panes of glass, with no two panes being the same size. To support the canopy, a structure was composed of strategically located "trees" that were carefully situated to minimize impact and obstruction. Balancing structural requirements with a visual lightness, each tree has a 2 m high tapered steel tubular trunk filled with heavily reinforced concrete. From the top of each primary trunk, three steel tubes branch out at different angles and then individually split into a quad of four smaller branches that connect to the roof mesh. The glass canopy, tubular branches and structural trees create a visual tableau that explores the potentiality and interrelation of the natural and man - made while establishing an undeniable sense of place.

Designed as a connector to weave central Warsaw's urban fabric back together, Zlote Tarasy includes an interior circulation path that connects to the city's existing pedestrian patterns, linking to the Warszawa Centralna train station on the south with pocket parks and the historic city center on the north. Additional entrances around the perimeter of the project help recreate the historic urban grid lost during World War II and revitalize public spaces nearby. Situated within the master plan area for the Palace of Culture, Zlote Tarasy is expected to play an integral role in connecting the Palace with a new high-rise district planned for the city.

**Location**
Warsaw, Poland

**Area**
186,000 sq.m.

**Main materials**
Glass, concrete, stone,

**Company**
The Jerde Partnership

**Designer**
The Jerde Partnership, Epstein Sp. z.o.o, Arup, Waagner Biro, EDAW, Sussman/Prezja, Mace Polska

**Photographer**
Marcin Czajkowski

屋顶由4,700个三角玻璃窗格构成，但是没有任何两块窗格是一样的尺度。为了支撑天篷，设计师设计了一个树形结构，对其精心定位以便将影响和阻碍最小化。用视觉亮度平衡结构要求，每棵树有一个2m高的锥形钢管主干，中间充斥着大量钢筋混凝土。3根钢管从每个主干的顶部按不同的角度伸展出来，然后逐个分裂成4个较小的分支，与屋顶的网格相连。玻璃天篷，管状分支以及结构树创造出一幅生动的视觉盛宴，挖掘出自然物与人造物品的潜力和相互联系，同时还建立了一种不可否认的地方意识。

本案作为一个编织中央华沙城市网的连接器，Zlote Tarasy加入了一条与城市现有人行道模式相连的室内环形小道，将南部带有一个袖珍花园的华沙Centralna火车站和北部历史悠久的城市中心连接起来。围绕项目周围增建的入口利于重建在二战中瘫痪的具有重大历史意义的城市电网，还有助于复兴周边的公共空间。坐落在文化宫的总体规划区内，人们期待Zlote Tarasy能在连接文化宫和为城市规划的新高楼区之间发挥重要的作用。

ice cream cafe
APART
poziom 0

poziom 3
ZŁOTE TARASY
poziom 3
jak trafić na parking
BURGER KING

poziom 2
Złote
poziom 2

buduar
PALMERS

SPHINX

WAYNE'S COFFEE

SABRA
REPLAY

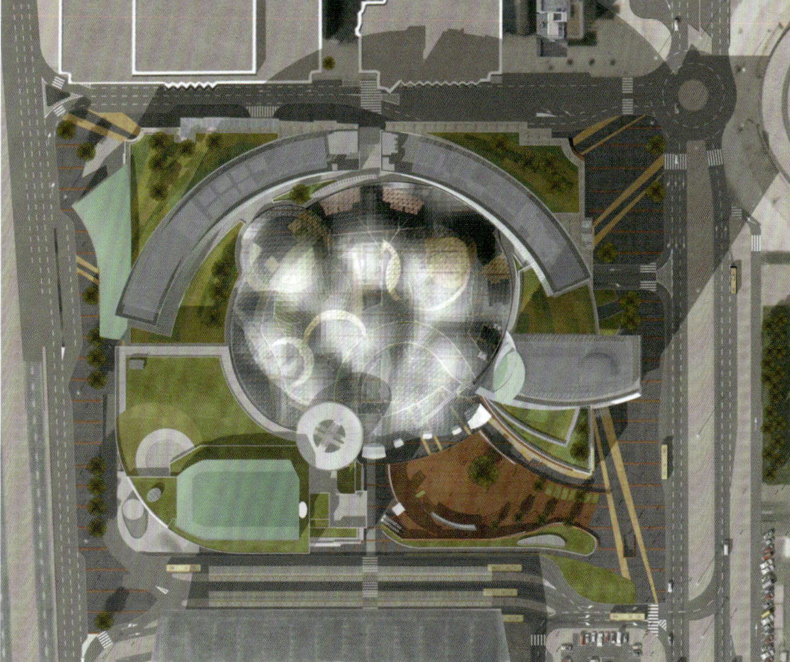

# Changwon City 7
# 昌原城市7号

The City 7 Mall's main element is a canal that animates the project as it links its different areas. From its origin in the rooftop park, water travels through canals and waterfalls as it makes its way down through the restaurants and shops to the hotel and convention center. At the lower levels, water takes a variety of forms, from a rapid stream to placid reflective areas to interactive fountains, to create a number of experiences. Landscaping, with plants featuring a vivid color palette and variety of fragrances, ties The City 7 Mall together from the richly planted canal and throughout its many districts.

Trees placed around the perimeter of the project help visually establish The City 7 Mall as the new heart of the surrounding district, while enhancing the pedestrian environment of the overall area. Providing a human - scale canopy, the trees create a distinctive natural edge for the project and establish the prototype for future enhancement of surrounding streets. The City 7 Mall is planned to create a singular pedestrian experience that connects the nearby convention center and residential communities to the project's hotel, shops, restaurants, offices, and parks.

## Location
Changwon, South Korea

## Area
422,441 sq.m.

## Company
The Jerde Partnership

## Designer
The Jerde Partnership, Executive Architect & Residential/Office Architect: EAWES Architects

## Photographer
Dosi Saram

城市7号购物中心的主要元素是一条运河，它连接了项目的不同区域，因而激活了项目。从屋顶花园开始，水柱穿过运河、瀑布，然后再顺流而下经过餐馆和商店，直到酒店和会议中心。在较低楼层，水流形式变化多端，从一股湍流变为平静的反射区，再到互动式喷泉，创造出多种体验。建筑内的绿色植物以生动的调色板和各种香料植物为特色，如此美丽的风景从种满水草的运河到其经过的许多区域环绕着整个城市7号商场。

项目周边种植的树木在视觉上将城市7号购物中心建立成周边地区的核心，同时还提升了这个地区的人行道环境。这些树还提供了一个近景树冠的风景，打造出项目独有的自然优势，为将来提升周边街区的环境，树立了一个原型。设计师计划在城市7号购物中心建造一个独特的人行道，将附近的会议中心和住宅区与该项目的酒店、商店、餐厅、办公室和公园相连接。

FRANCK PROVOST

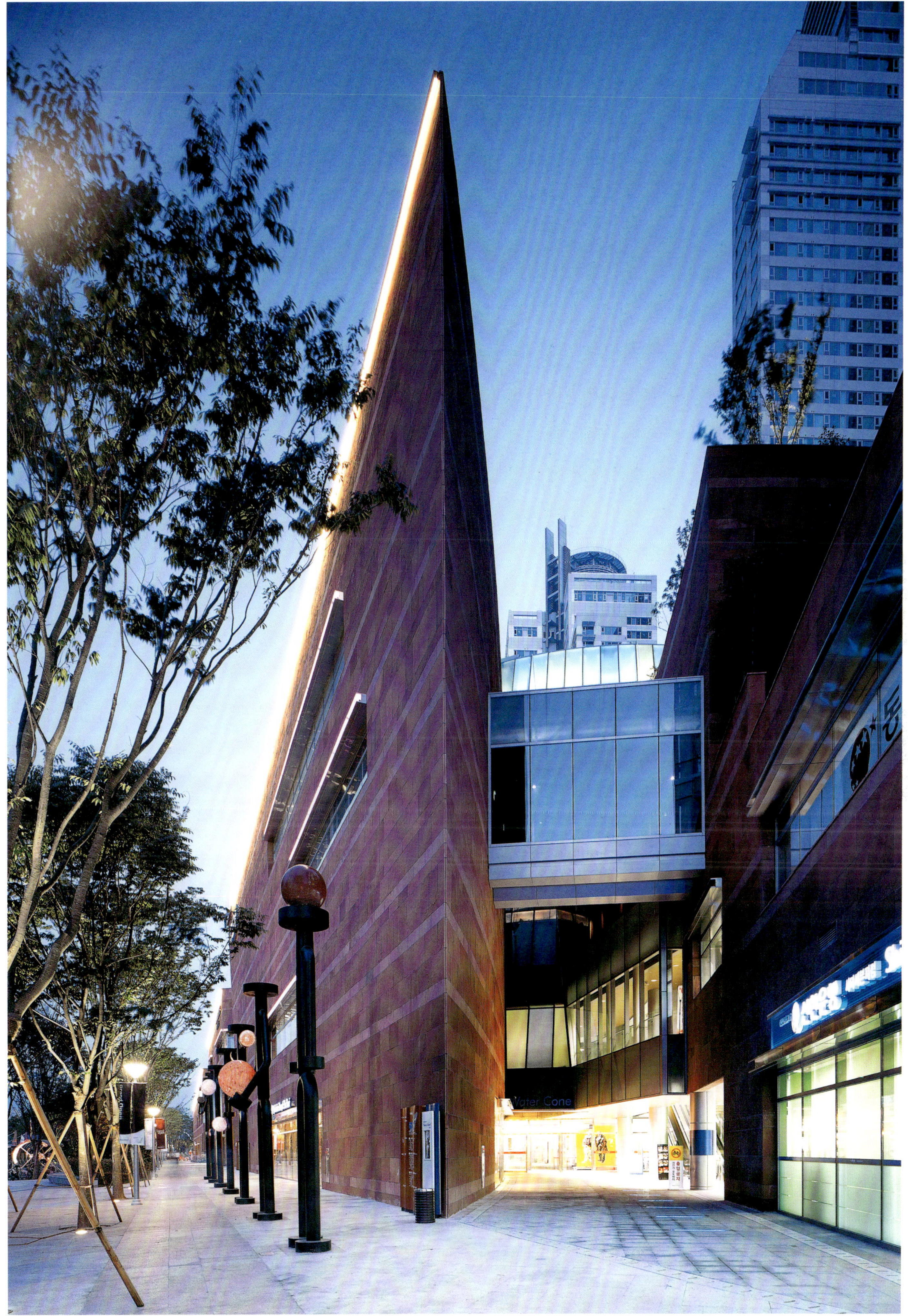

E-113
BR robbins

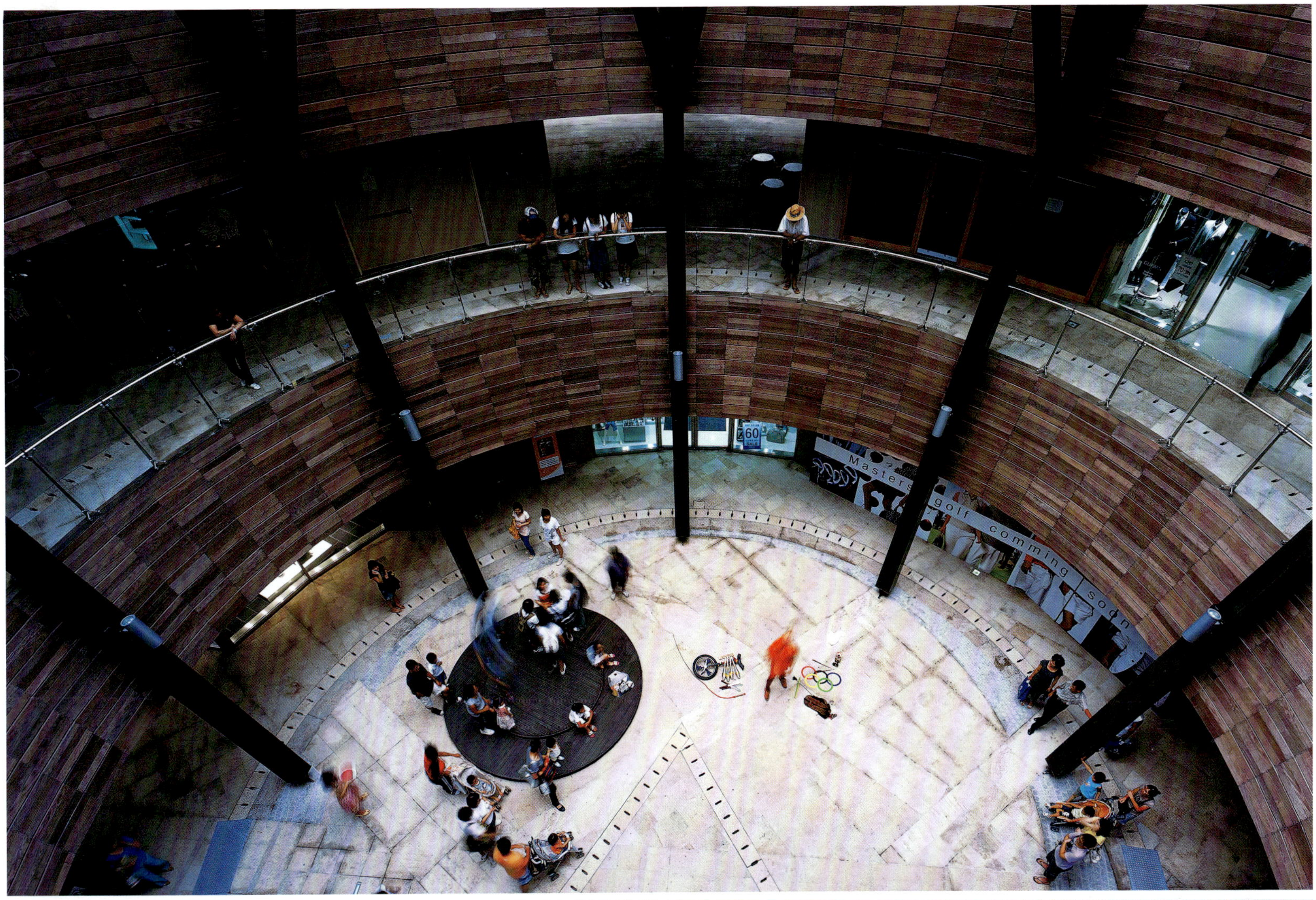
Masters golf comming soon

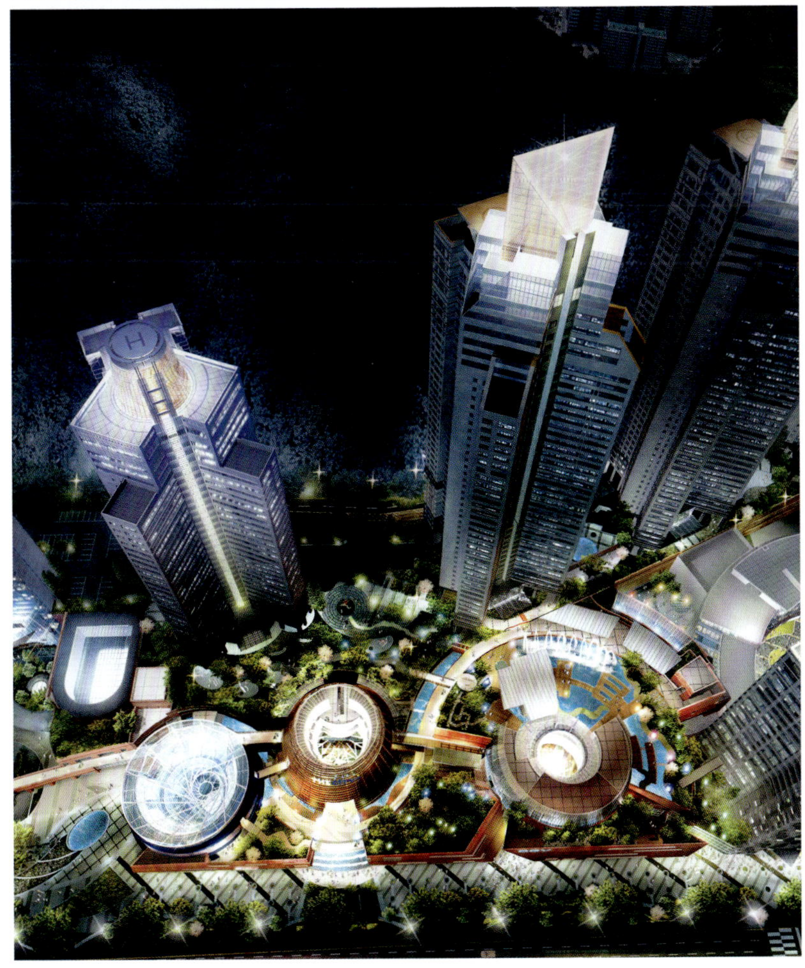

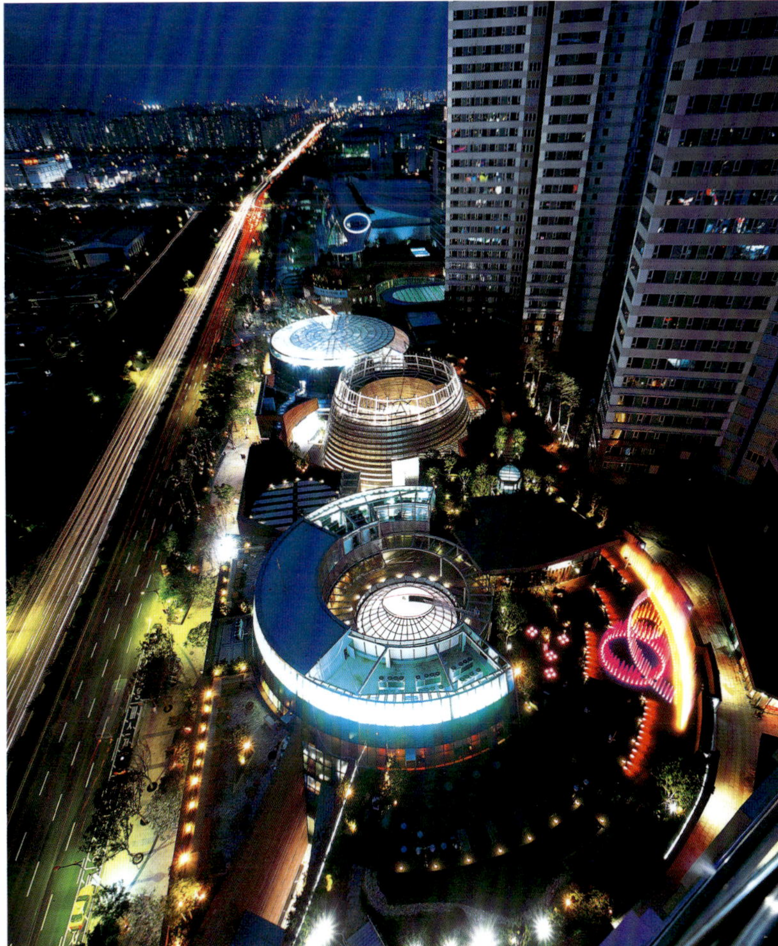

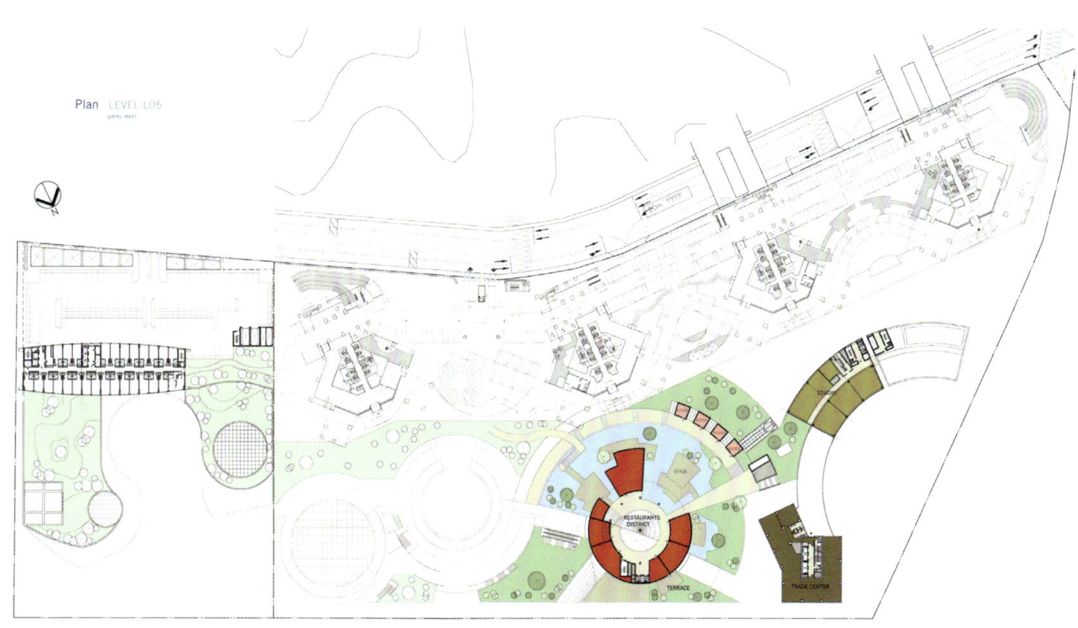
Plan LEVEL L05
RESTAURANTS DISTRICT
TERRACE
TRADE CENTER

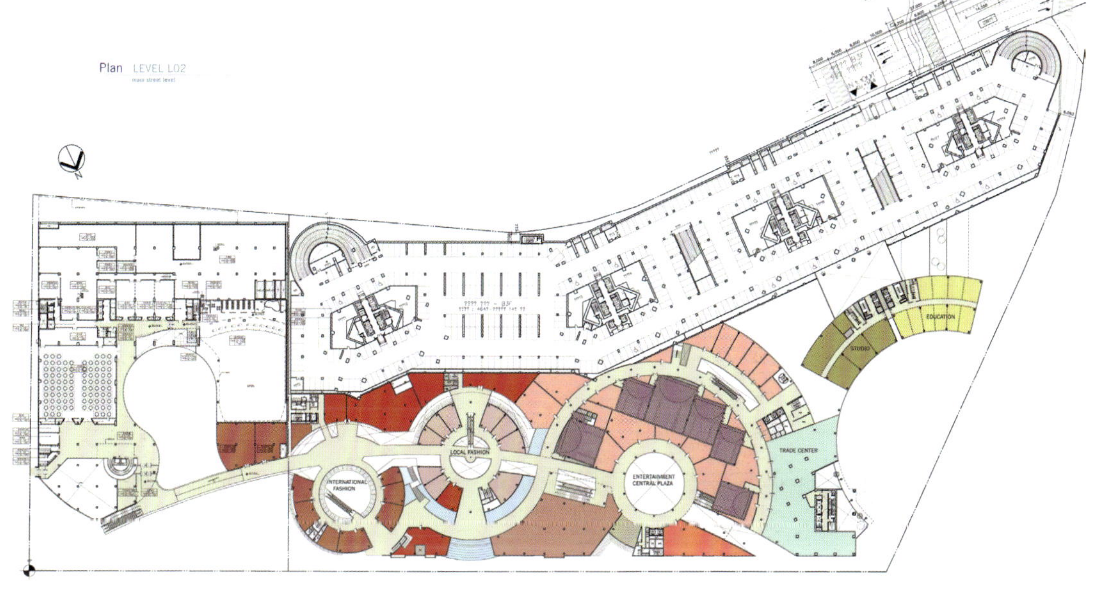
Plan LEVEL L02
EDUCATION
STUDIO
LOCAL FASHION
INTERNATIONAL FASHION
ENTERTAINMENT CENTRAL PLAZA
TRADE CENTER

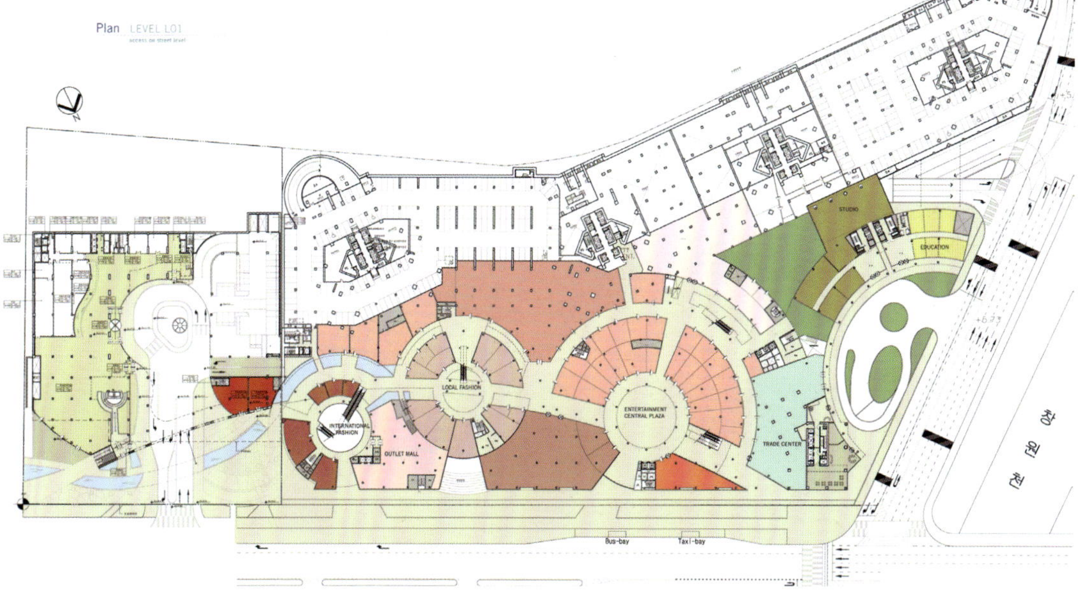
Plan LEVEL L01
STUDIO
EDUCATION
LOCAL FASHION
INTERNATIONAL FASHION
OUTLET MALL
ENTERTAINMENT CENTRAL PLAZA
TRADE CENTER
Bus-bay
Taxi-bay

# Namba Parks
# 难波公园

The project is located on a 3.7 hectare site of the former Nankai Stadium, a prime redevelopment location in Osaka. The project is surrounded by raised railroad tracks to the east and an urban boulevard and elevated viaduct to the west.

Namba Parks, a vibrant lifestyle center, inserts a much - needed natural amenity into Osaka's dense city core. Namba Parks generously weaves rich landscaping and other natural elements with specialty retail, entertainment and dining, creating a new unique place that celebrates the interaction of people, culture and recreation. Located adjacent to Namba Train Station, the first stop on the new line connecting the city to Kansai Airport; Namba Parks is a key gateway project for Osaka that will redefine the city's identity and urban experience.

Cascading down from the eighth - level rooftop is another Jerde innovation: a series of green terraces atop the roofs of the retail spaces below. Extending the canyon theme to the roof, Jerde brings the canyon - top landscape to its very precipice. Forming 1.15 hectares (2.8 acres) of rooftop park space, one of Japan's largest, this "Big Park" is a counterpart to the "Big City" contained below in Namba Parks' retail spaces. The roof park features trees, miniature ponds, shrubbery and planting beds—all irrigated by recycled water filtered from the graywater of the restaurants within the complex. During the summer, when asphalt can reach a surface temperature of 51 degrees Celsius (124 degrees Fahrenheit) and concrete is 45 (113), the rooftop park is only 34 (93).

**Location**
Osaka, Japan

**Area**
130,064 sq.m.

**Main materials**
Concrete, stucco,

**Company**
The Jerde Partnership

**Designer**
Nikken Sekkei, Obayashi Corporation

**Photographer**
Hiroyuki Kawano

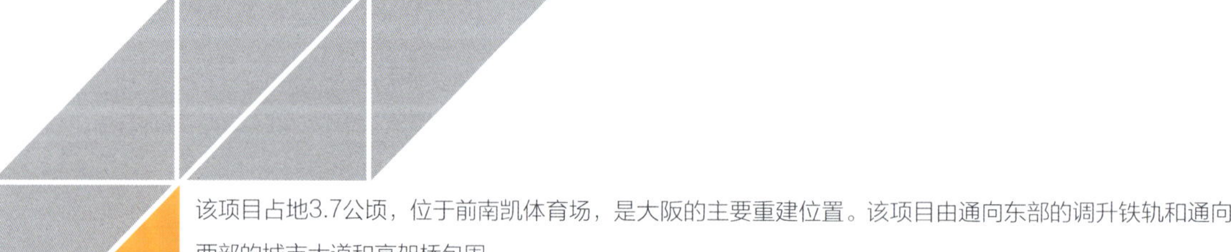

该项目占地3.7公顷，位于前南凯体育场，是大阪的主要重建位置。该项目由通向东部的调升铁轨和通向西部的城市大道和高架桥包围。

难波公园，一个充满活力的生活中心，给大阪密集的城市核心注入了急需的自然美化要素。难波公园将丰富的园林绿化等自然要素与专业零售、娱乐及餐饮慷慨地编织在一起，创建一个庆祝人民、文化和娱乐互动的独特的新地方。难波公园毗邻新线路的第一站——难波火车站，新线路的第一站连接了城市和关西机场；难波公园是是大阪的主要门户项目，将重新定义该城市的身份和城市经验。

从第八层屋顶向下的级联是Jerde的另一个创新：在下面零售空间的屋顶之上是一系列绿色露台。Jerde将峡谷的主题扩展到屋顶，将顶部峡谷景观引入悬崖。形成1.15公顷（2.8亩）的屋顶公园空间，是日本最大的屋顶公园空间之一，这个“大公园”是包含在“难波公园”的零售空间下面的“大城市”的对应物。屋顶公园有树木、微型池塘、灌木、花圃，全部由再生水灌溉，这个再生水是由此综合建筑内餐厅的可再利用废水过滤而来。在夏季，沥青表面的温度可51℃（124℉），水泥可达45℃（113℉），而屋顶公园只有34℃（93℉）。

Alfa Romeo
Alfa Romeo

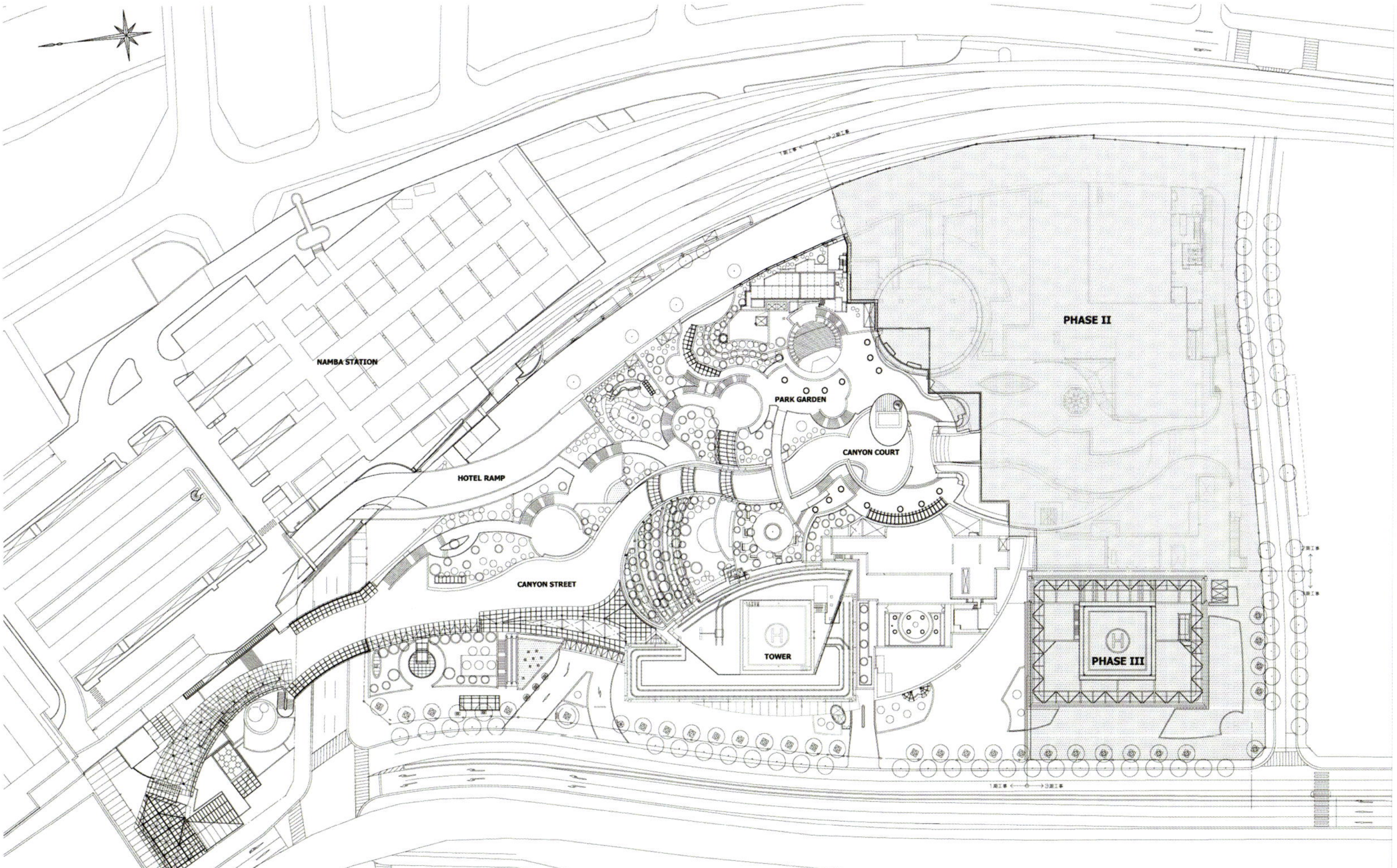

NAMBA STATION
HOTEL RAMP
CANYON STREET
PARK GARDEN
CANYON COURT
TOWER
PHASE II
PHASE III

SITE PLAN

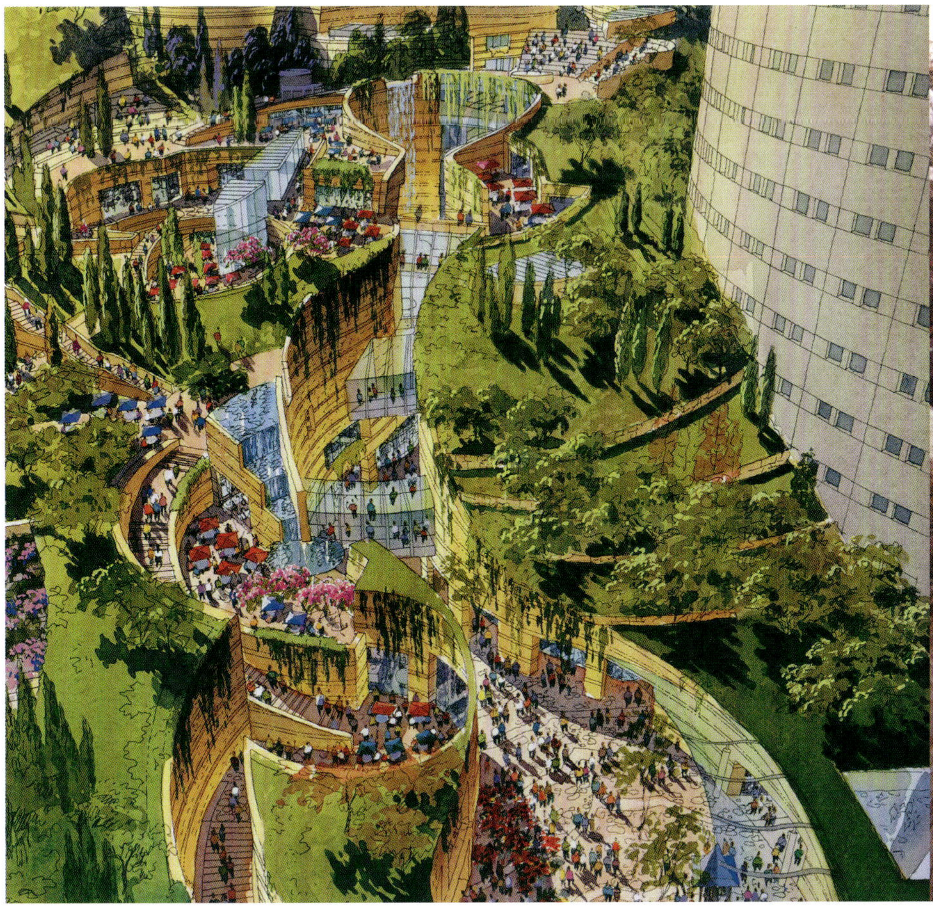

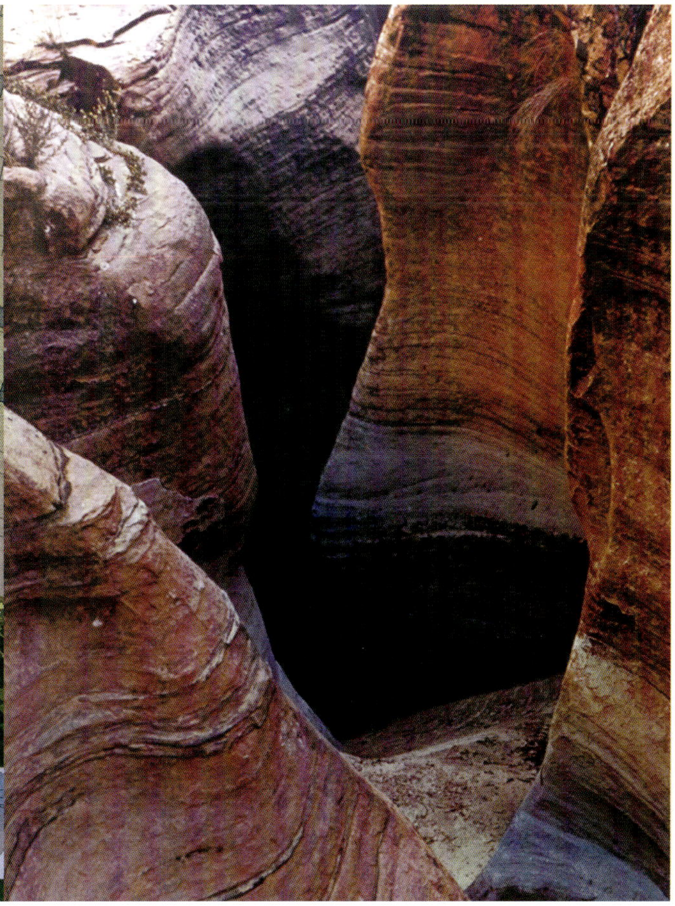

# Galleria Centercity

## Galleria Centercity购物中心

The Galleria Cheonan building organisation employs a propeller principle. Four stacked programme zones, each thematically combining three storeys and containing public plateaus, are linked to the central void. The concept of the propeller is a fluent upstream flow of people through the building, whilst the propeller wings simultaneously stream visitors outwards to the plateaus on the various levels. On the facades a gradual transition from exterior surface to the interior plateaus accentuates the internal organisation. The facade is double layered. Both the outer glass shell and the inner skin comprise of a linear pattern created by the vertical mullions. During the day the building has a monochrome reflective appearance, whilst at night soft colours are used to generate waves of coloured light across the facade.

天安购物中心的建筑团体采用了螺旋桨的原理。每个区域主题上都结合了3层楼并包括公共露台的四叠方案的区域与中央空隙连接。螺旋桨的概念是人们通往此建筑的一个流畅的上游流，而螺旋桨翼同时使外面的访客涌入各层的露台。在外墙，从外表面到内部露台的逐步过渡突出了内部安排。建筑物的外表是双层的。外层的玻璃壳和内壳板由垂直竖框创建的一个线性模式组成。白天此建筑有一个单色反射外观，而在夜间柔和的色彩用于产生贯穿整个表面的彩色光波。

**Location**
Cheonan, Korea

**Area**
66,700 sq.m.

**Main materials**
Flooring: wood, composite tiling; Glazing: toughened glass with safety foils and decoration foil; Ceilings: gypsum board; Wall finishes: gypum board, mdf paneling; Lighting: light slots with translucent plastic.

**Company**
UN Studio

**Designer**
Ben van Berkel

**Photographer**
Christina Richters and Kim Yong-kwan

809
046
Card Center
카드센터

# Printemps Haussman Department Store
## 春天百货

The re-design of Printemps, completed in September of 2010, combines the Paris tradition with a fresh contemporary environment. With an overall plan of replacing the typical "department store" image, Yabu Pushelberg combines freshness and energy in subtle, yet elegant ways. Utilizing fun and quirky fixtures, Yabu Pushelberg focused on creating a promenade - type space, with a unifying framework strong enough to provide spacial definition, but also open and transparent enough to be enticing and inviting.

With the use of fine finishes and materials throughout, Printemps' retail space is conceived as a series of "rooms", like a large mansion, each with its own unique and identifiable character, creating an image that exudes a chic flavor in the eyes of visitors and an international influence in those of the Parisian.

春天百货的重新设计于2010年9月竣工，结合了巴黎的传统与清新的现代环境。因有着取代典型的“百货公司”形象的总体规划，设计师以含蓄而优雅的方式将新颖和活力结合在一起。设计师利用千奇百怪的装饰着重创造一个长廊式的空间，并带有统一的框架，不但强大到足以界定空间，而且开放、通透到足以让人沉醉其中。

由于精致饰面和材料的使用贯穿整个空间，春天百货的零售空间被设想为一系列的“房间”，如同一座大厦，每个房间都有其独具一格的特色，营造了一个既能给游客带来别致风情又能给巴黎人带来国际影响力的印象。

**Location**
Paris, France

**Area**
8,999.69 sq.m.

**Main materials**
Display cases and fixtures, glass screen, curve wall panel

**Company**
Yabu Pushelberg

**Designers**
George Yabu, Glenn Pushelberg

**Photographer**
Richard Powers

PRADA
Dior
SWAROVSKI

CELINE


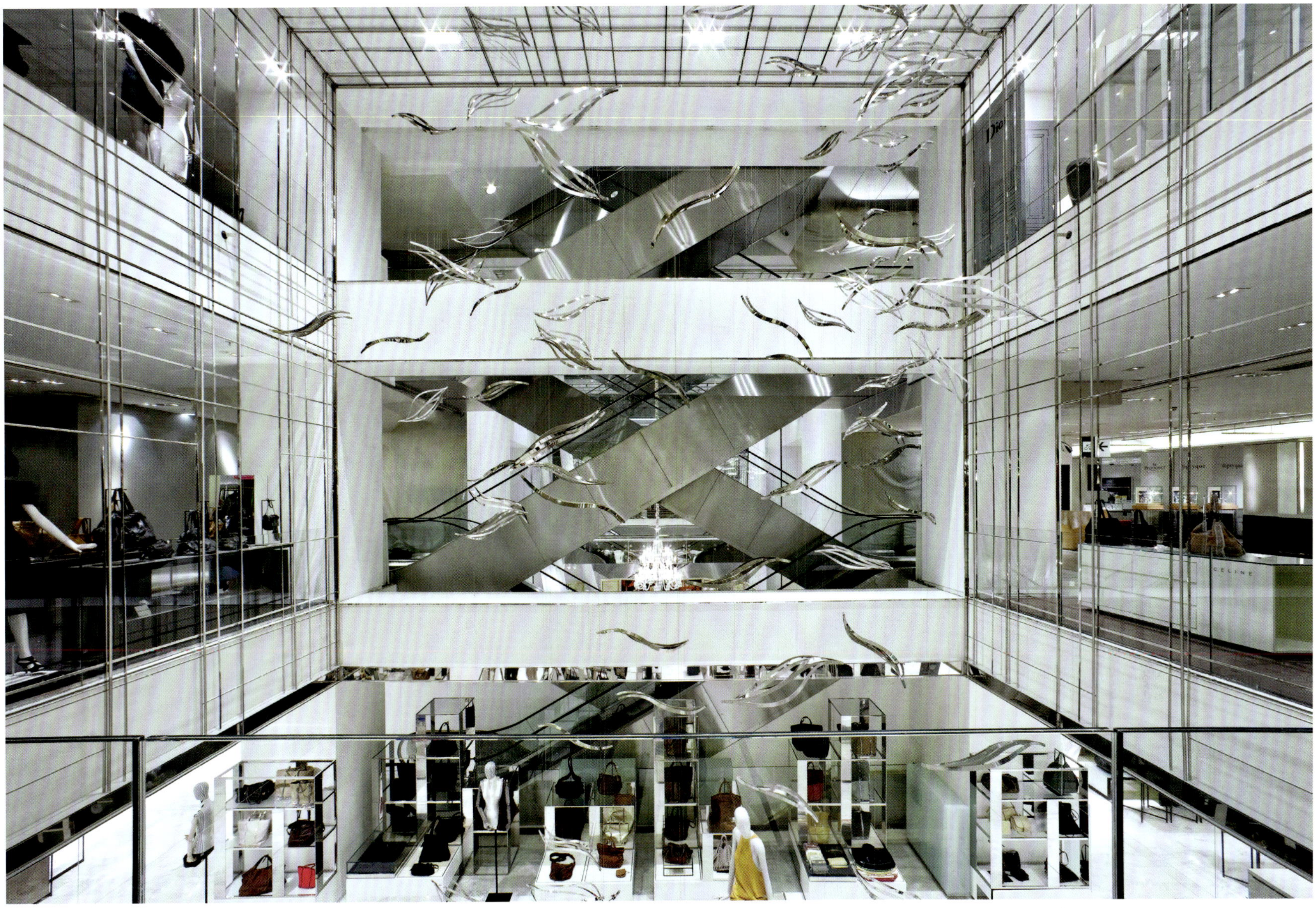

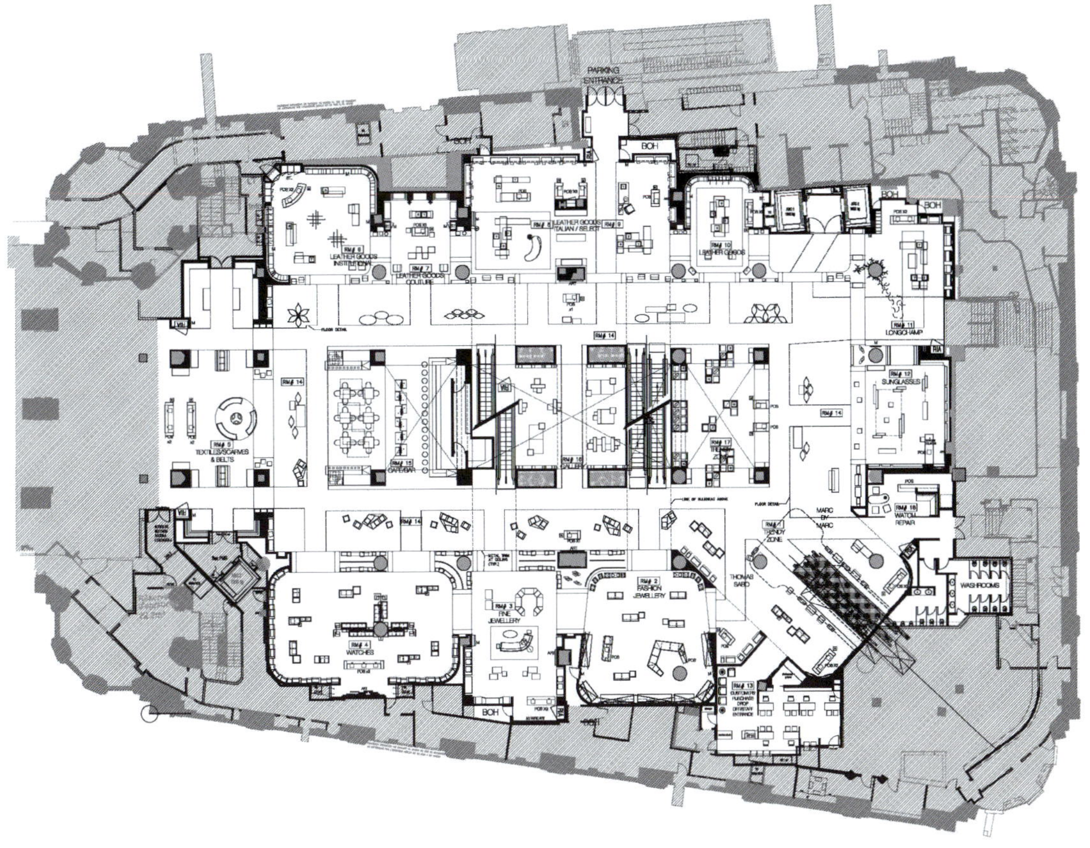

PARKING ENTRANCE
BOH
LEATHER GOODS INSTITUTIONAL
LEATHER GOODS COUTURE
LEATHER GOODS ITALIAN / SELECT
LEATHER GOODS
LONGCHAMP
SUNGLASSES
TEXTILES/SCARVES & BELTS
CAFE/BAR
GALLERY
WATCH REPAIR
MARC BY MARC
TRENDY ZONE
THOMAS SABO
WASHROOMS
FASHION JEWELLERY
FINE JEWELLERY
WATCHES
CUSTOMER PURCHASE DROP OFF/STAFF ENTRANCE

FURLA

BURBERRY
BALENCIAGA

LADURÉE

GUESS

RADO
JAEGER-LECOULTRE
OMEGA
BVLGARI
LONGINES

IWC
SCHAFFHAUSEN
BREITLING

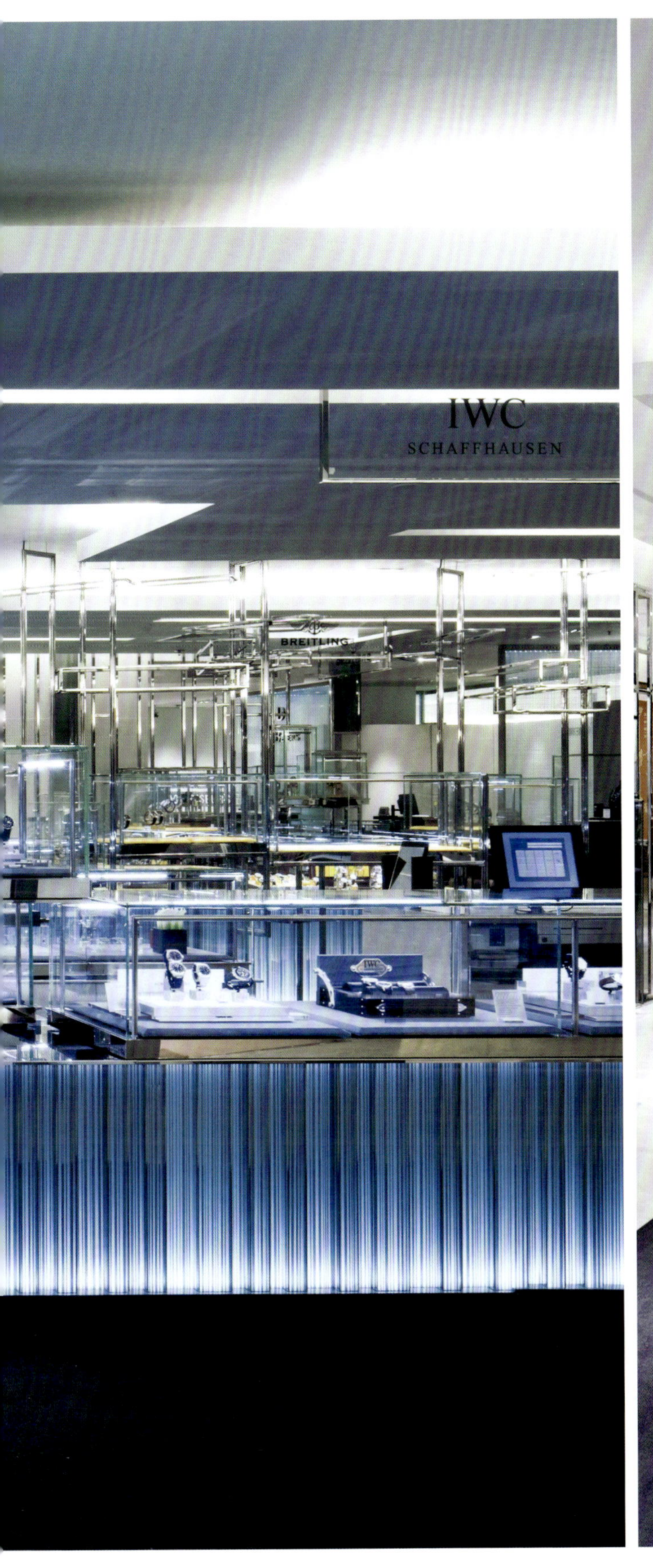

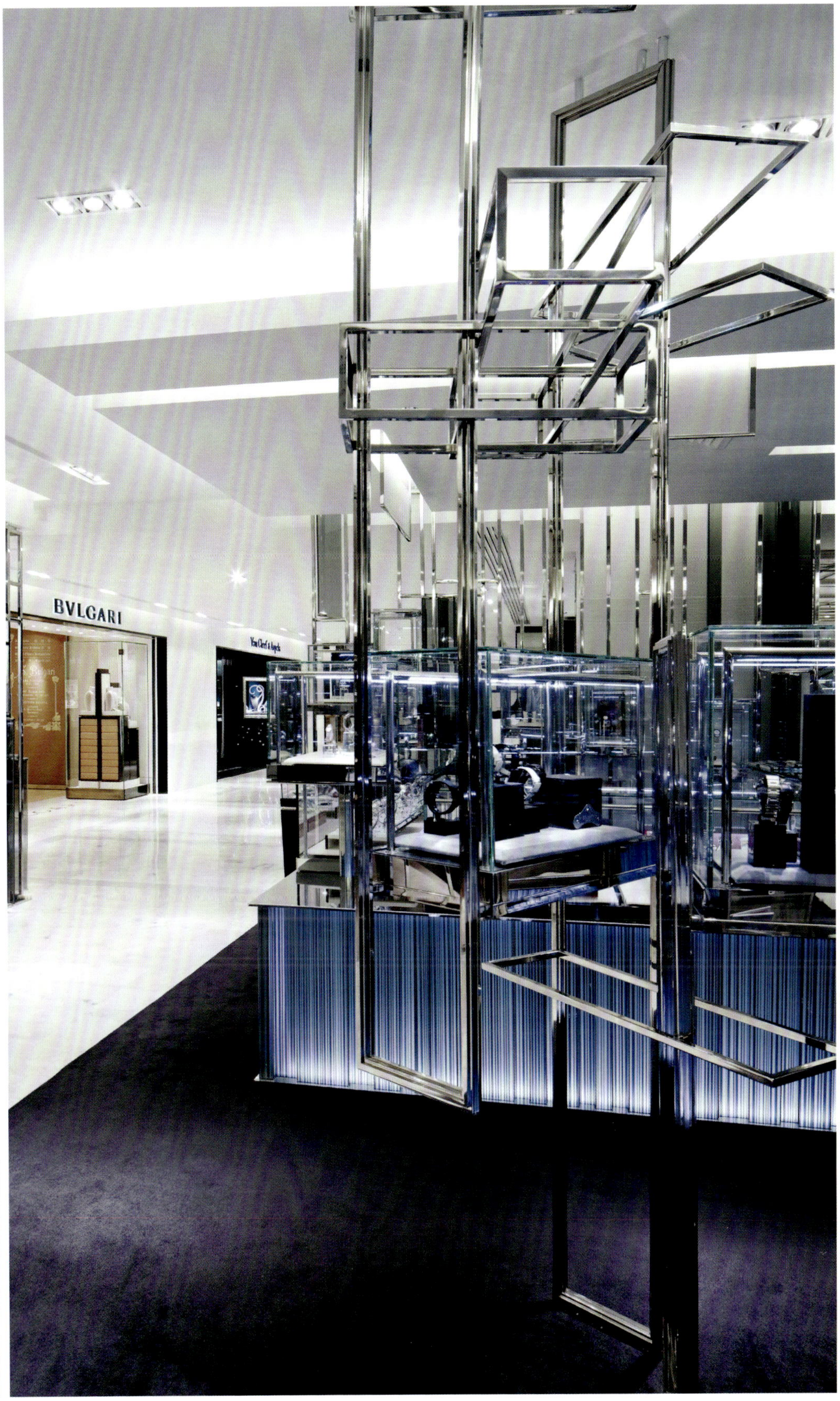
BVLGARI

# FRIIS Aalborg City Centre
# 弗里斯奥尔堡城市中心商场

The new building and the former department store are located on opposite sides of Nytorv, the main in central Aalborg, but are linked by two footbridges, one of which is broad enough to provide space for shops, which enjoy a view of the city life and a newly - built square.

The shops are located at street level and on the first floor. A unifying element is the 25 m tall central space in the new part of the Centre. The space extends from the lowest parking deck to the uppermost ceiling, where escalators and elevators mark the Centre's pulse.

The aim has been to unite the entire complex in a single, easily readable building volume which binds the various functions together while at the same time emphasizing the city's frontage. The building unites its many different functions with a white stucco facade and striking glass panels in structural glazing, but the facade also reflects its diversity, just as it reflects the diverse structure of the surrounding city - from the large - scale open areas towards the harbor to the denser, more intimate shopping streets.

**Location**
Aalborg, Denmark

**Area**
67,000 sq.m.

**Main Materials**
Pre-fab Concrete, Perforated steel cladding, insulated render facades

**Company**
C. F. Møller Architects

**Designer**
C. F. Møller Architects

**Photographer**
Helene Høyer Mikkelsen, Steen & Strøm

大楼和之前的百货公司分别坐落在中部奥尔堡的主要干道Nytory的两侧，但是由两条天桥相连接，其中一条十分宽阔，足够提供打造出商店的空间，顾客在此可俯瞰到都市生活和一座新建的广场。

该店面位于街道层面第一层。统一元素就是一个位于商场新建部分的25m高的中央空间。空间从最低的停车甲板延伸到最上面的天花板，其间自动扶梯和电梯成为该购物中心的脉搏。

设计的目的就是用一个单一的、易辨认的建筑体融合整个综合体，将各种功能区联系在一起，同时还突出城市的临街面。该建筑通过一面白色的粉刷外墙和结构玻璃上的显著的玻璃板块连接不同的功能，但是，正如外墙反映出周边城市的多样性结构一样，外墙也反映出该建筑的多样性，从面朝海港的大型开放性区域，到更密集、更贴心的购物街。

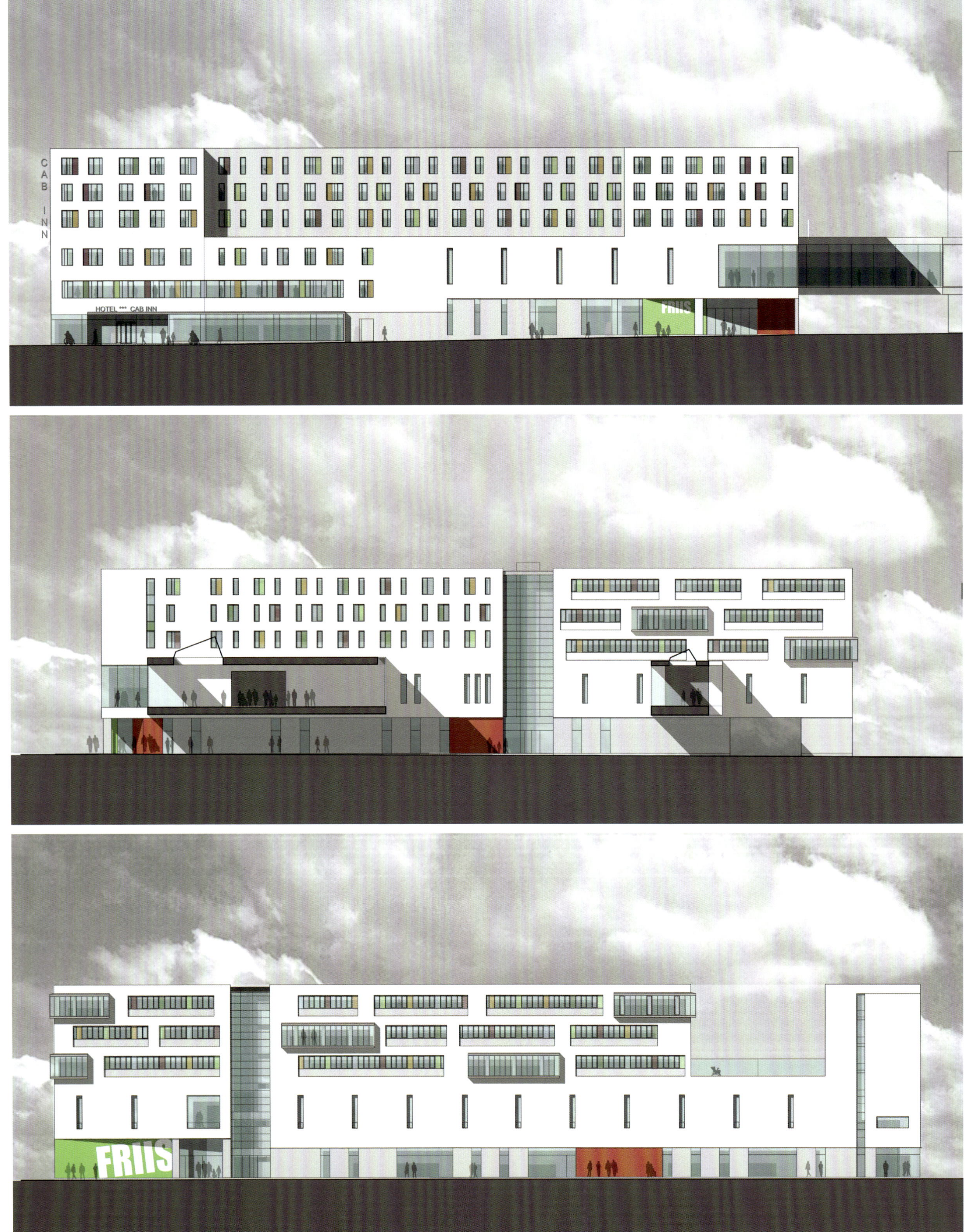
CAB INN
HOTEL *** CAB INN
FRIIS
FRIIS

Friis
H&M

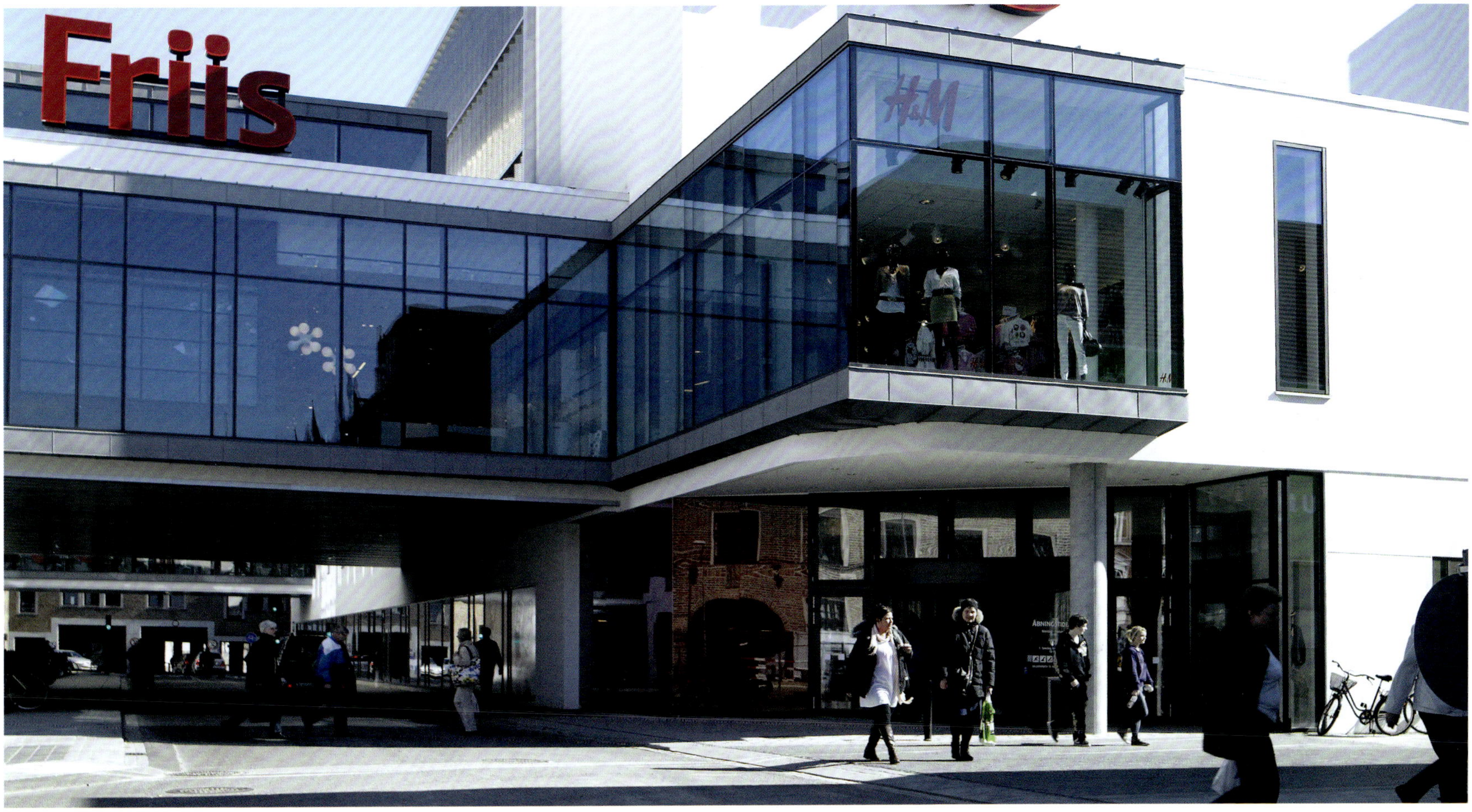

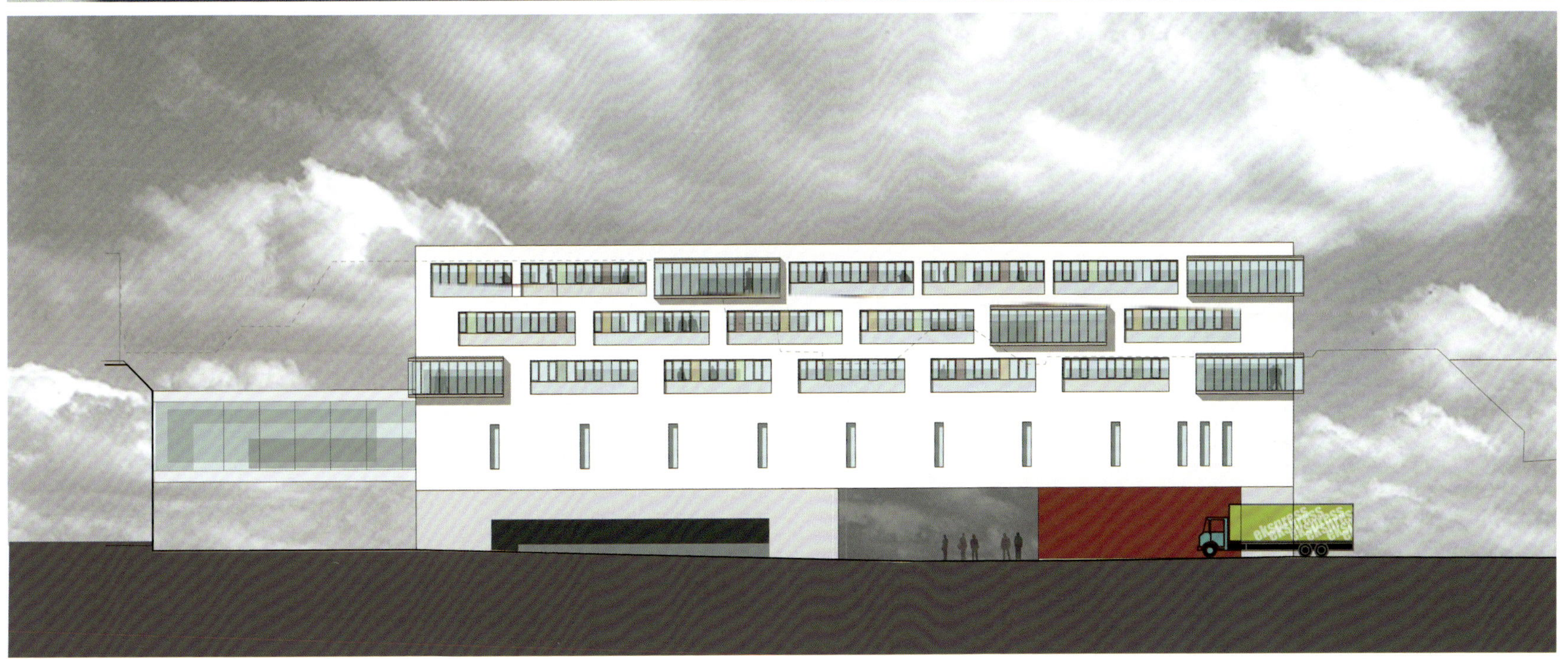

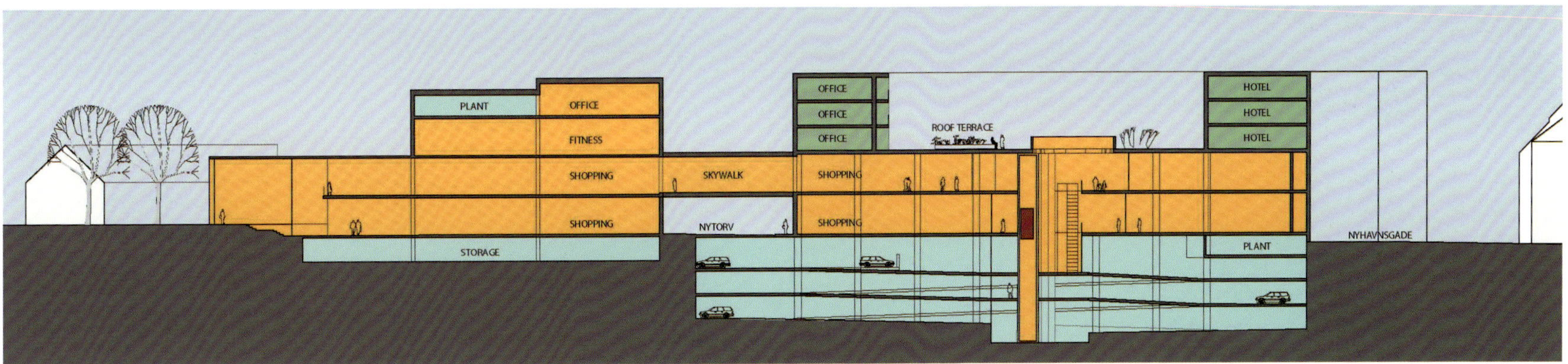

PLANT
OFFICE
FITNESS
SHOPPING
SHOPPING
STORAGE
SKYWALK
NYTORV
OFFICE
OFFICE
OFFICE
ROOF TERRACE
SHOPPING
SHOPPING
HOTEL
HOTEL
HOTEL
PLANT
NYHAVNSGADE

Nyhavnsgade
HOTEL
SHOPS
Fjordgade
PASSAGE
DELIVERIES
SHOPS
Rendsburggade
Nytorv
SHOPS
PASSAGE
SHOPS

FRIIS
MODERN
STEAKHOUSE

PANDORA

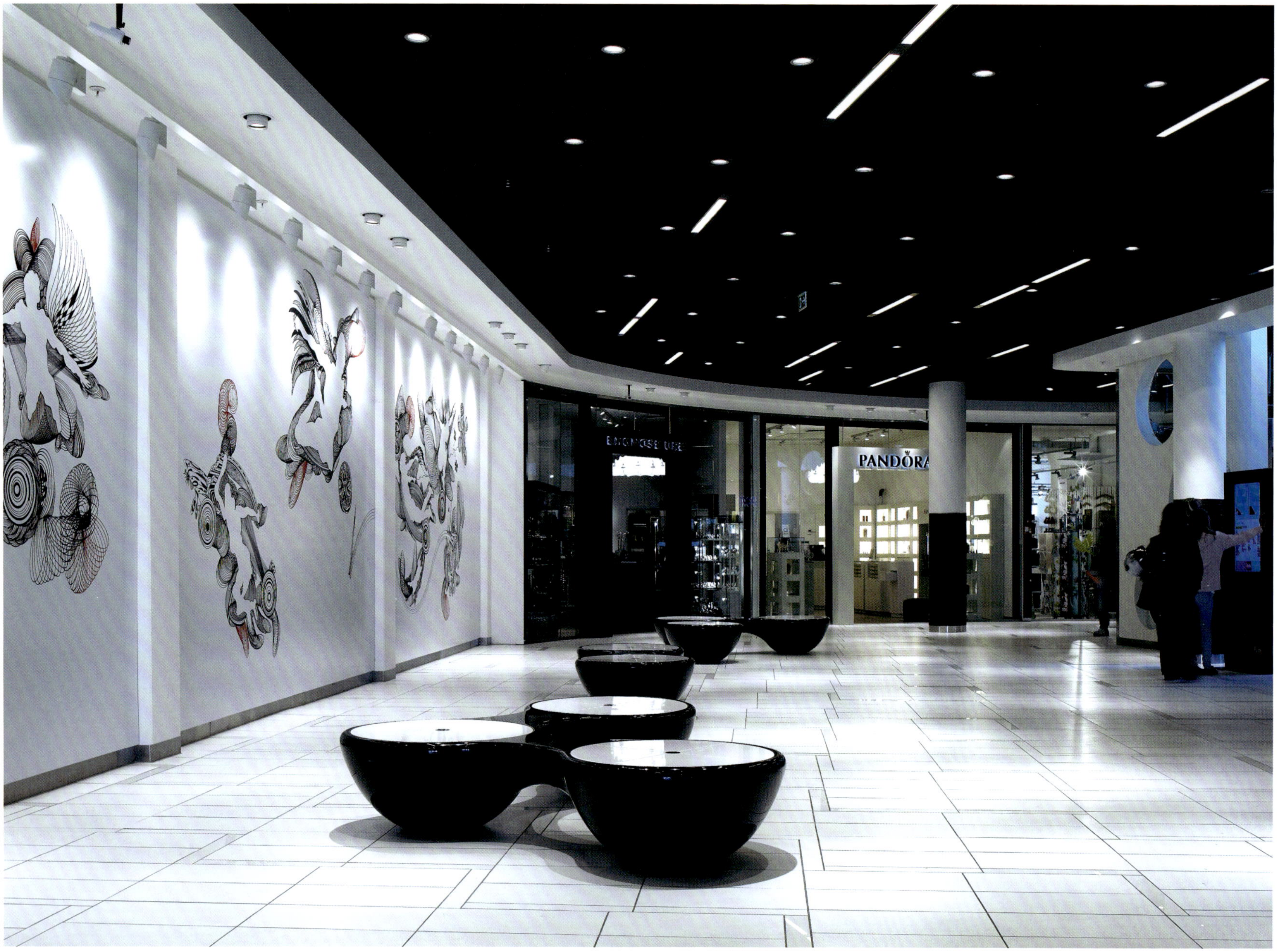

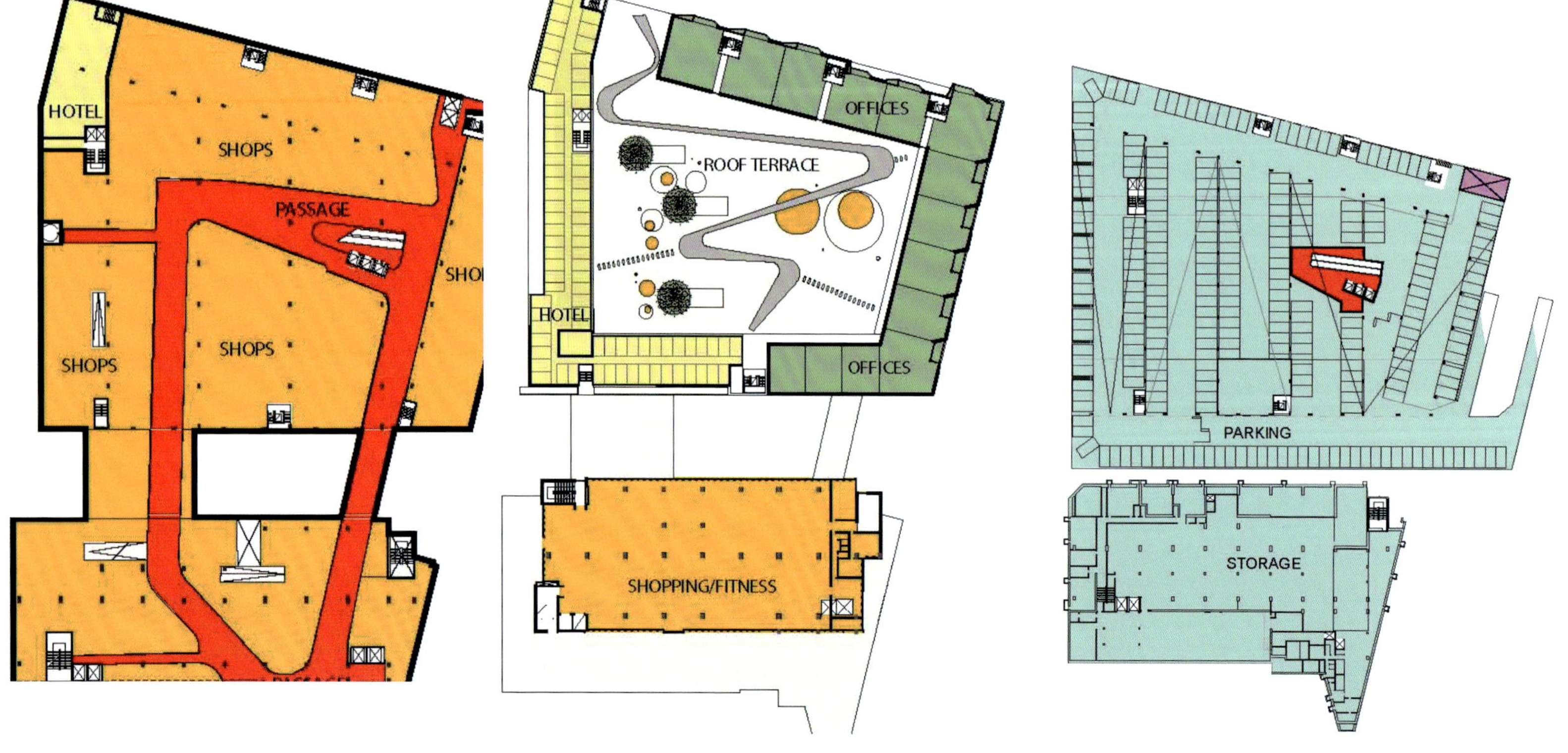

HOTEL
SHOPS
PASSAGE
SHOPS
SHOPS

OFFICES
ROOF TERRACE
HOTEL
OFFICES
SHOPPING/FITNESS

PARKING
STORAGE

# The Gateway

## The Gateway商场

Salt Lake City had lost commercial activity, particularly retail, and housing to the suburbs. In 1997, Jerde identified the site as an opportunity to bring business and activity back to the downtown core and teamed up with a local developer, The Boyer Company, to transform the abandoned rail yard into a community focal point.

The challenge was to generate a vibrant, urban center for Salt Lake City with a strong pedestrian identity, within a brown field site. The goal was to also ignite further development around the site's three city blocks.

The 141,000 square meters of buildings have a consistent design that is a celebration of the natural setting of Salt Lake City, incorporating brick, sandstone and concrete. The project is not enclosed, but open to the air and the outdoors. The restored, historic Union Pacific depot serves as the entry to The Gateway, opening to Olympic Legacy Plaza, a communal gathering place. An internal street connects the plaza to the project's south side, an area specifically devoted to nightlife and entertainment. The street is lined by two levels of retail, entertainment and cultural uses. Housing is placed above retail on the west side of the plaza with office space above retail on the east side.

**Location**
Salt Lake City, Utah

**Area**
141,000 sq.m.

**Main Materials**
Limestone, glass, concrete

**Company**
The Jerde Partnership

**Designer**
The Jerde Partnership

**Photographer**
Michael McRae

盐湖城已经失去了商业活动，特别是零售店和郊区的住房。1997年，Jerde 将其视为一个可以为市中心带来商业和活动的机会，并和当地的一个发展商Boyer公司一起，将这个废弃的铁路院改造成一个社区焦点。

设计面临的挑战就是在一片褐地上，为盐湖城建造一个充满活力的城市中心，同时还具有较强的行人标志。设计的目标还在于促进围绕该址的3个城市街区的进一步发展。

这个141,000㎡的建筑群有一个始终如一的设计，其中包括砖块、砂岩和混凝土的运用，就是颂扬盐湖城的自然环境。该项目不是封闭的，而是向空气和户外开放。历史悠久的Union Pacific 工厂，重新装修后成为Gateway 的入口，面向公共聚会场所——奥运遗产广场。一条内侧街道将一个广场与建筑项目的南端相连，而南端是专门的夜生活娱乐区，街道两旁是两层的零售店、娱乐区和文化区。该项目就坐落在位于广场西侧的零售店之上，而商场东侧则是位于零售店之上的办公空间。

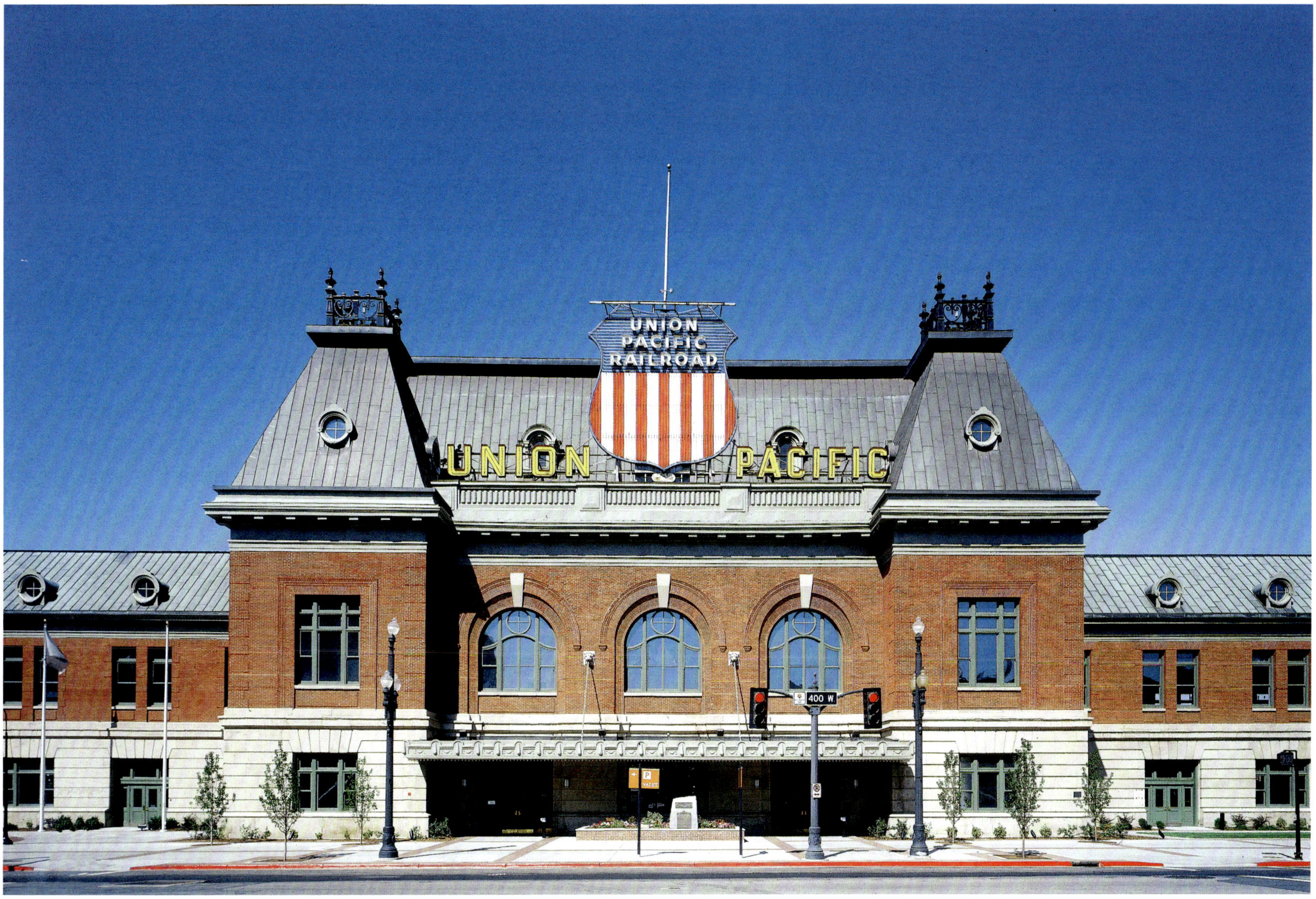
UNION
PACIFIC
RAILROAD
UNION PACIFIC
400 W

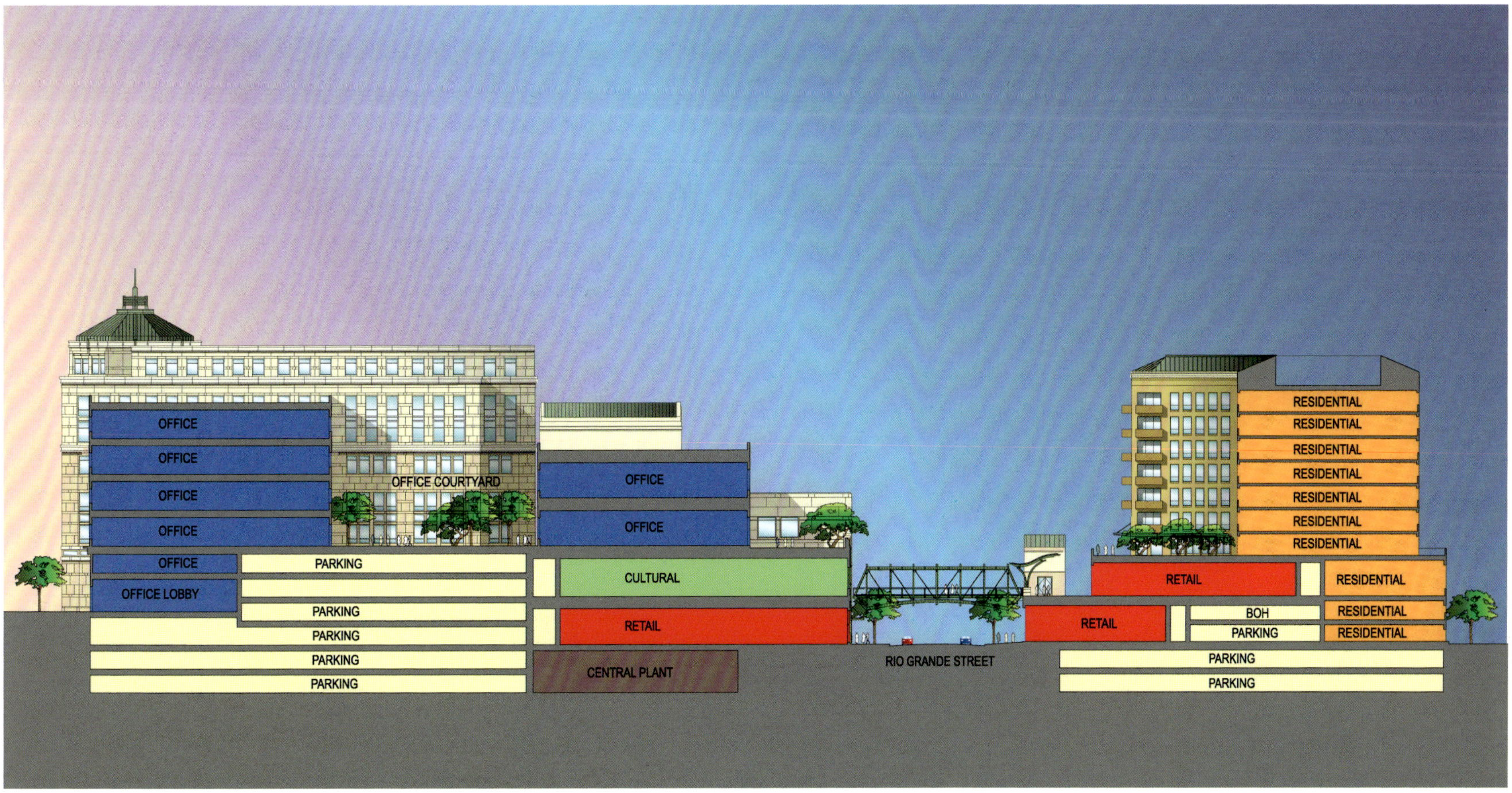
OFFICE
OFFICE
OFFICE
OFFICE
OFFICE
OFFICE LOBBY
OFFICE COURTYARD
PARKING
PARKING
PARKING
PARKING
PARKING
OFFICE
OFFICE
CULTURAL
RETAIL
CENTRAL PLANT
RIO GRANDE STREET
RESIDENTIAL
RESIDENTIAL
RESIDENTIAL
RESIDENTIAL
RESIDENTIAL
RESIDENTIAL
RESIDENTIAL
RETAIL
RESIDENTIAL
RETAIL
BOH
RESIDENTIAL
PARKING
RESIDENTIAL
PARKING
PARKING

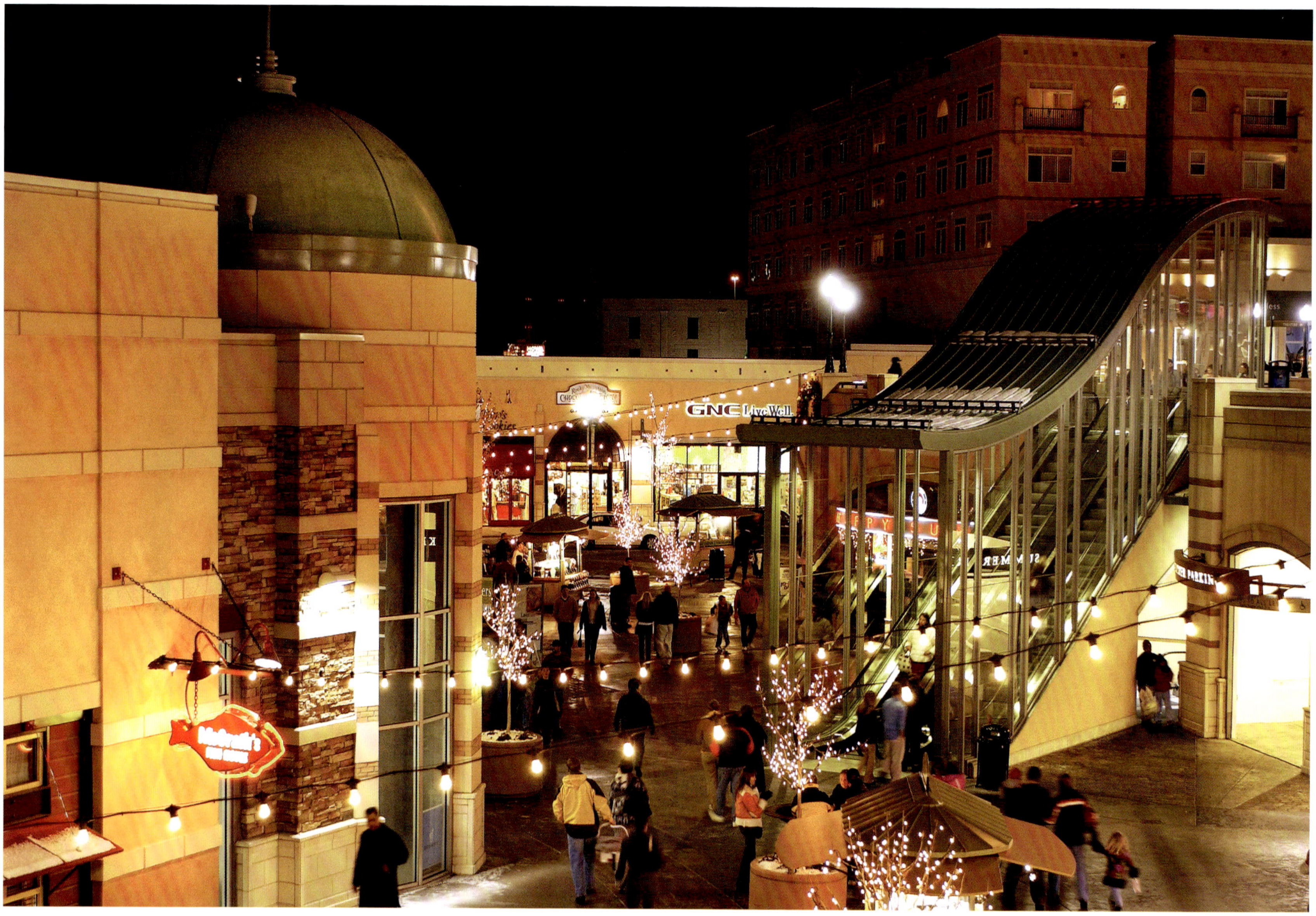
GNC Live Well

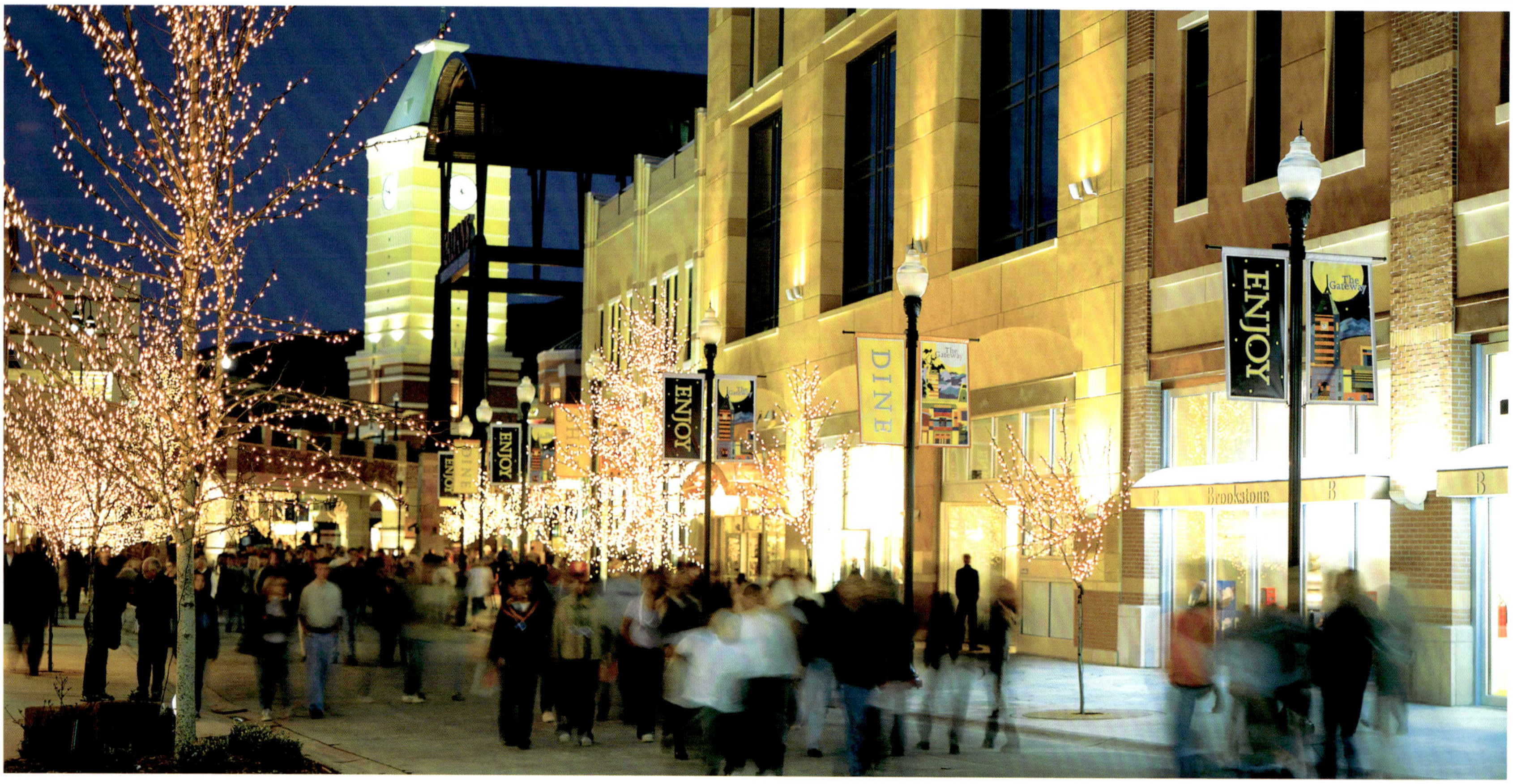
ENJOY
The Gateway
DINE
Brookstone

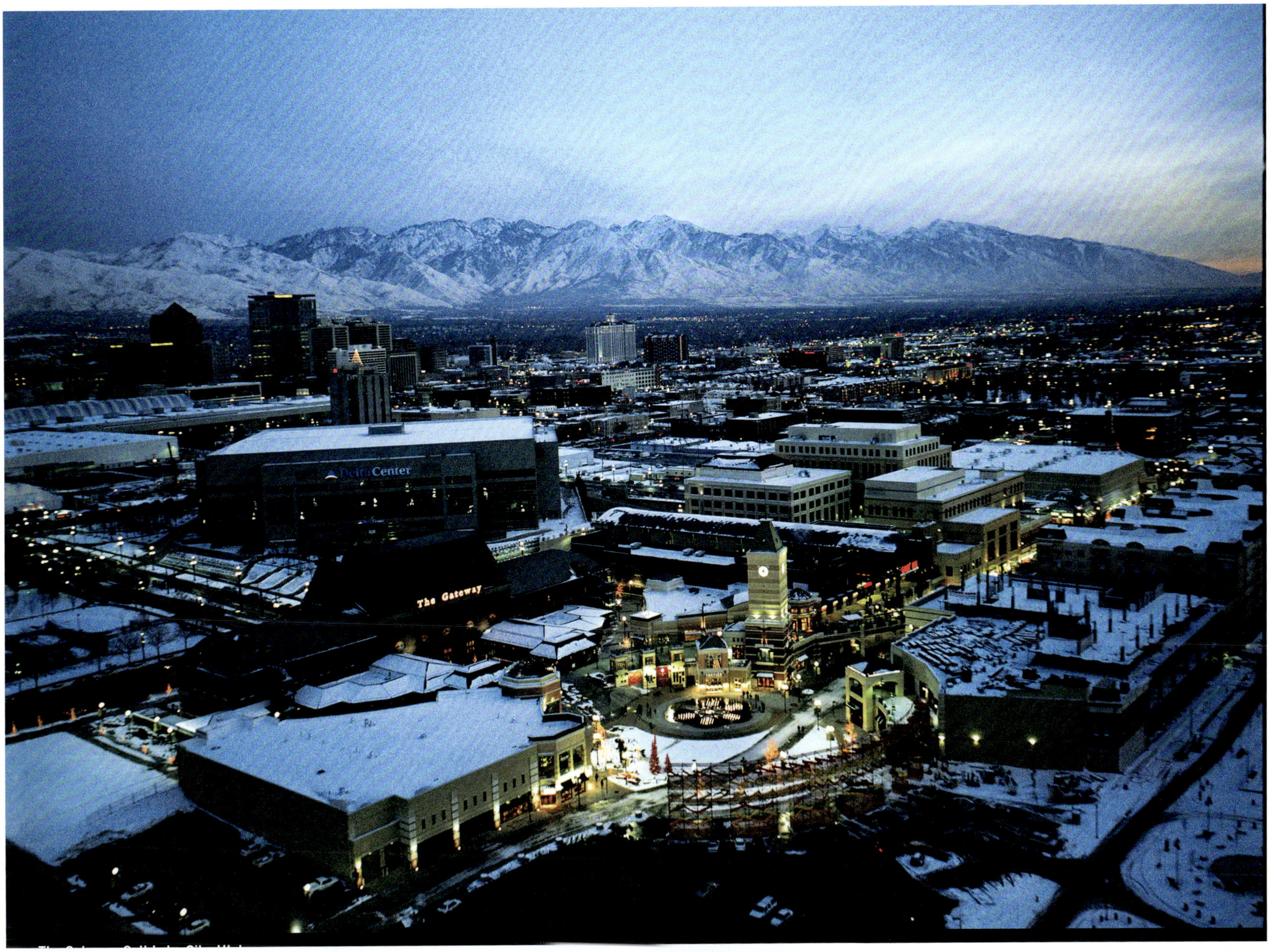

The Gateway
DINE
ENJOY
SHOP
ENJOY

Jazz

BARNES & NOBLE

# Roppongi Hills
# 六本木之丘

Jerde recognized that Roppongi Hills' location at the crossroads of the best of Tokyo - the business districts of Shimbashi and Toranomon; the commercial and trading markets of Aoyama and Akasaka; and the exclusive residential neighborhoods of Azabu and Hiroo - presented a unique opportunity to raise the bar for modern urban planning and bring lasting social, cultural and economic value to the community.

Roppongi Hills achieves an often discussion but seldom realized combination of dense mixed-use buildings – office, cultural, hotel, entertainment, retail and residential – with generous open spaces and parks. A natural, ungraded circulation composed of unique pathways that are lined with pedestrian-scaled experiences connects Roppongi Hills' range of uses and spaces.

Roppongi Hills includes many innovative features, such as a rooftop garden with rice paddies that also serves as a protective measure against earthquakes, state - of - the - art computer/information technology systems, double - decker elevators, and an environmentally friendly energy center with a gas-powered electric cogeneration system that can assure a 72 - hour supply of electricity in a blackout.

**Location**
Tokyo, Japan

**Area**
724,600 sq.m.

**Company**
The Jerde Partnership

**Designer**
The Jerde Partnership

**Photographer**
Daici Ano

Jerde认识到六本木之丘位于东京最好的十字路口——新桥和虎门商务区、青山和赤坂的商贸市场，麻布和广伟的专属居民区，为现代城市规划中提升酒吧以及为该社区带来持久的社会、文化和经济价值呈现独特的机会。

六本木之丘经常进行讨论，但却很少构建出华丽开放的空间和公园的密集型混合建筑的结合，例如办公室、文化区、酒店、娱乐中心、零售店和住宅区的结合。与人行道规模一致的特色小径组成了一条自然无序的圆环，将六本木之丘一系列的用途和空间连接起来。

六本木之丘拥有许多创新特色，例如，带有稻田的屋顶花园也包含了抗击地震的防护措施，先进的计算机和信息技术系统，双层电梯和环保能源中心，环保能源中心安置了一个以天然气为动力的电动热电联供系统，能保证72小时供电。

TOKYO SKY DECK

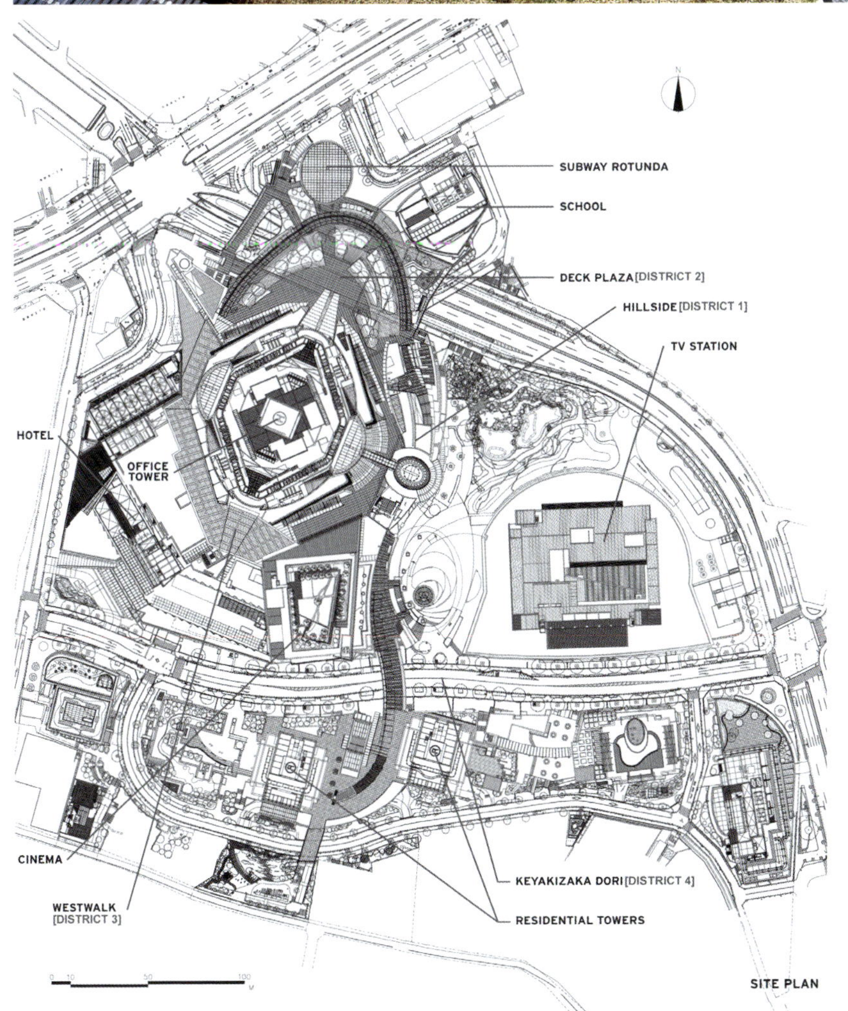
SUBWAY ROTUNDA
SCHOOL
DECK PLAZA [DISTRICT 2]
HILLSIDE [DISTRICT 1]
TV STATION
HOTEL
OFFICE TOWER
CINEMA
WESTWALK [DISTRICT 3]
KEYAKIZAKA DORI [DISTRICT 4]
RESIDENTIAL TOWERS
SITE PLAN

roppongi hills
roppongi hills
roppongi hills
roppongi hills
roppongi hills
roppongi hills

roppongi hills
roppongi hills
roppongi hills
roppongi hills

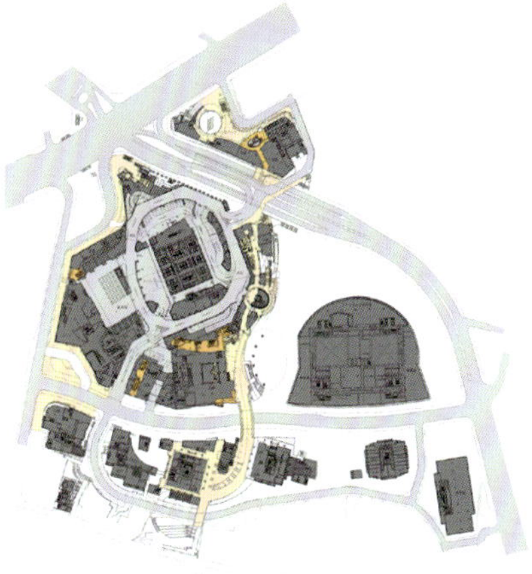

L01

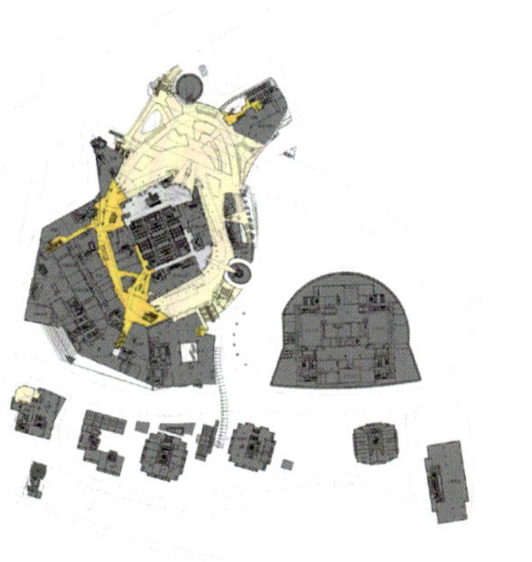

L02

B02

B01

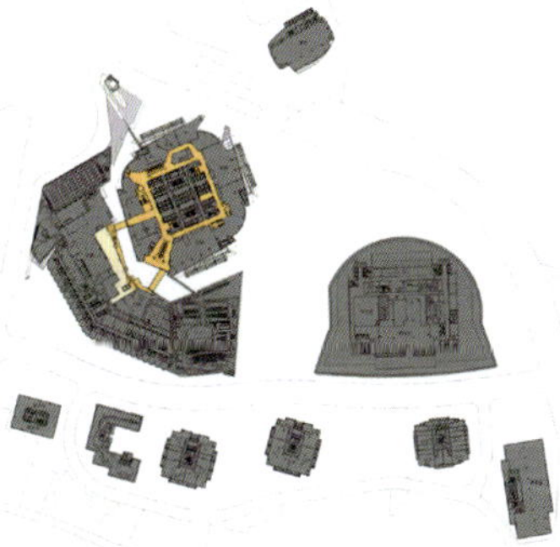

L05

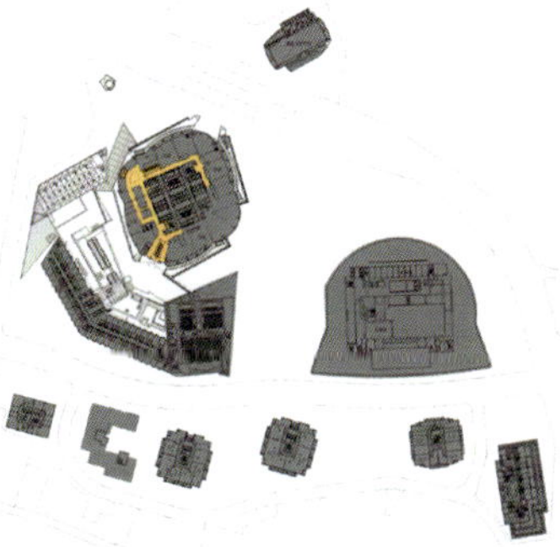

L06

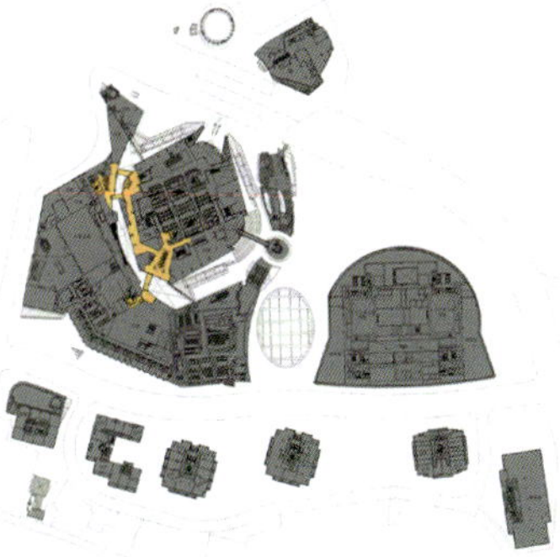

L03

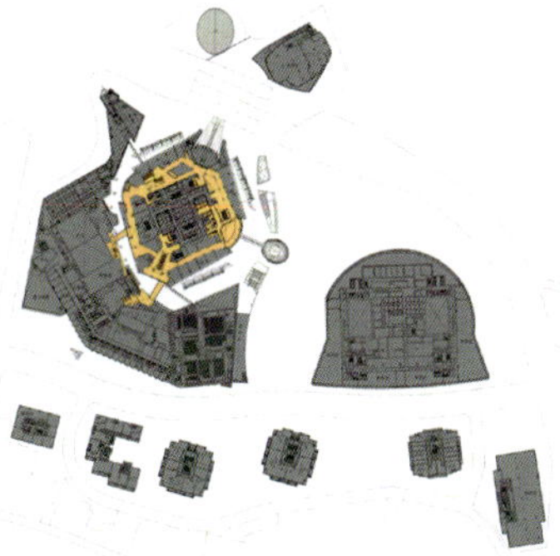

L04

六本木
マンション

EDGE

# Kanyon
# 大峡谷

Jerde designed Kanyon to be an open, vibrant and green project. The design is envisioned as a family of bold architectural shapes, each housing a different use, that come together to form a dynamic and iconic composition. Between the architectural shapes, the perspectives and views through the project constantly change, resulting in a vibrant and energetic space that engages users.

Kanyon's uses are connected by a central, open-air walkway that carves soft curves in the buildings' forms, creating a dramatic "canyon" effect. The 600 - foot - long canyon connects the office, residential, retail and entertainment uses with an exploratory path that features a variety of courtyards, terraces and other spaces where workers, residents and visitors to the project can interact. At the heart of the project, the canyon opens to a performance plaza, an amphitheater carved into the base of the project's entertainment sphere that will be the stage for activities of all sizes, from strolling entertainers to symphonies. Landscaping and water features enhance the project's open and natural design.

The "canyon" is lined on both sides by the project's four-level retail and entertainment complex. One side features three levels of retail and one level of restaurants, nightclubs and other entertainment venues. The four levels are terraced, allowing visitors on each level to overlook the activity occurring in the canyon below. The other side features the four - level entertainment sphere that contains a nine - screen cinema. The roof of the entertainment sphere features a restaurant with outdoor seating that enables diners to view the activity occurring within the project below.

**Location**
Istanbul, Turkey

**Area**
225,000 sq.m.

**Main Materials**
Concrete, glass, limestone, plaster, stone, metal

**Company**
The Jerde Partnership

**Designer**
The Jerde Partnership, Tabanlıo, lu Mimarlik, Arup

**Photographer**
Ali Kabas

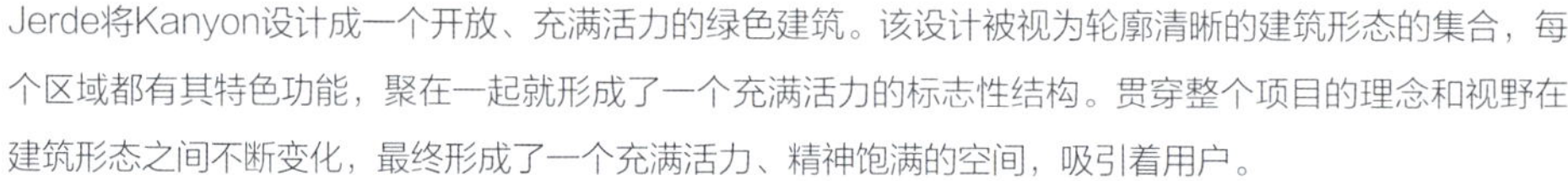

Jerde将Kanyon设计成一个开放、充满活力的绿色建筑。该设计被视为轮廓清晰的建筑形态的集合，每个区域都有其特色功能，聚在一起就形成了一个充满活力的标志性结构。贯穿整个项目的理念和视野在建筑形态之间不断变化，最终形成了一个充满活力、精神饱满的空间，吸引着用户。

一条核心的露天人行道在建筑的形式上雕刻出柔和的曲线，并将Kayon的各个功能联系起来，形成一种戏剧性的峡谷效应。这条600英尺长的峡谷通过一条非凡的小径将办公、住宅、零售和娱乐功能联系在一起。这条小径的特点是带有各种庭院、阳台和其他空间，供使用该项目的职员、居民和顾客进行互动。在项目中心，峡谷面向一个表演广场敞开，这个圆形剧场被雕刻成该项目娱乐区域的基底，为从漫步艺人到交响乐等各种规模的活动提供了舞台。绿化景观和特色水景加强了空间的开放和自然设计。

“峡谷”的标识排列在这个集零售商场和娱乐中心为一体的混合建筑体的两侧。其中一边以占地3层的零售商场、占地1层的餐厅、夜总会和其他娱乐场所为特色。4层楼呈梯形结构，方便顾客在峡谷的每一层都能俯瞰到下面其他楼层的活动。另一边则以占地四层的娱乐区为特色，其中包括一家有9个屏幕的剧院。娱乐区的顶层则以一个带有户外座椅的餐厅为特色，方便用餐者看到商场内下面楼层的活动。

NetWork
kanyon

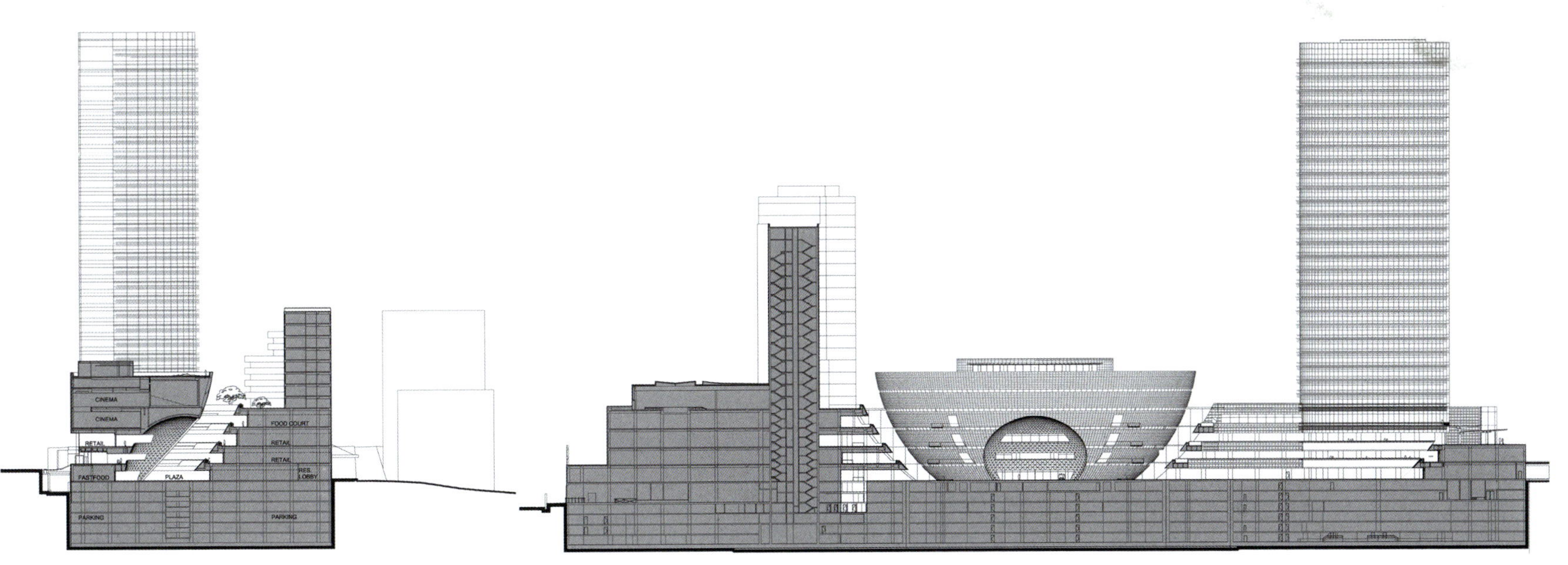
CINEMA
CINEMA
RETAIL
FOOD COURT
RETAIL
RETAIL
FASTFOOD
PLAZA
RES.
LOBBY
PARKING
PARKING

DERMOD
unitim

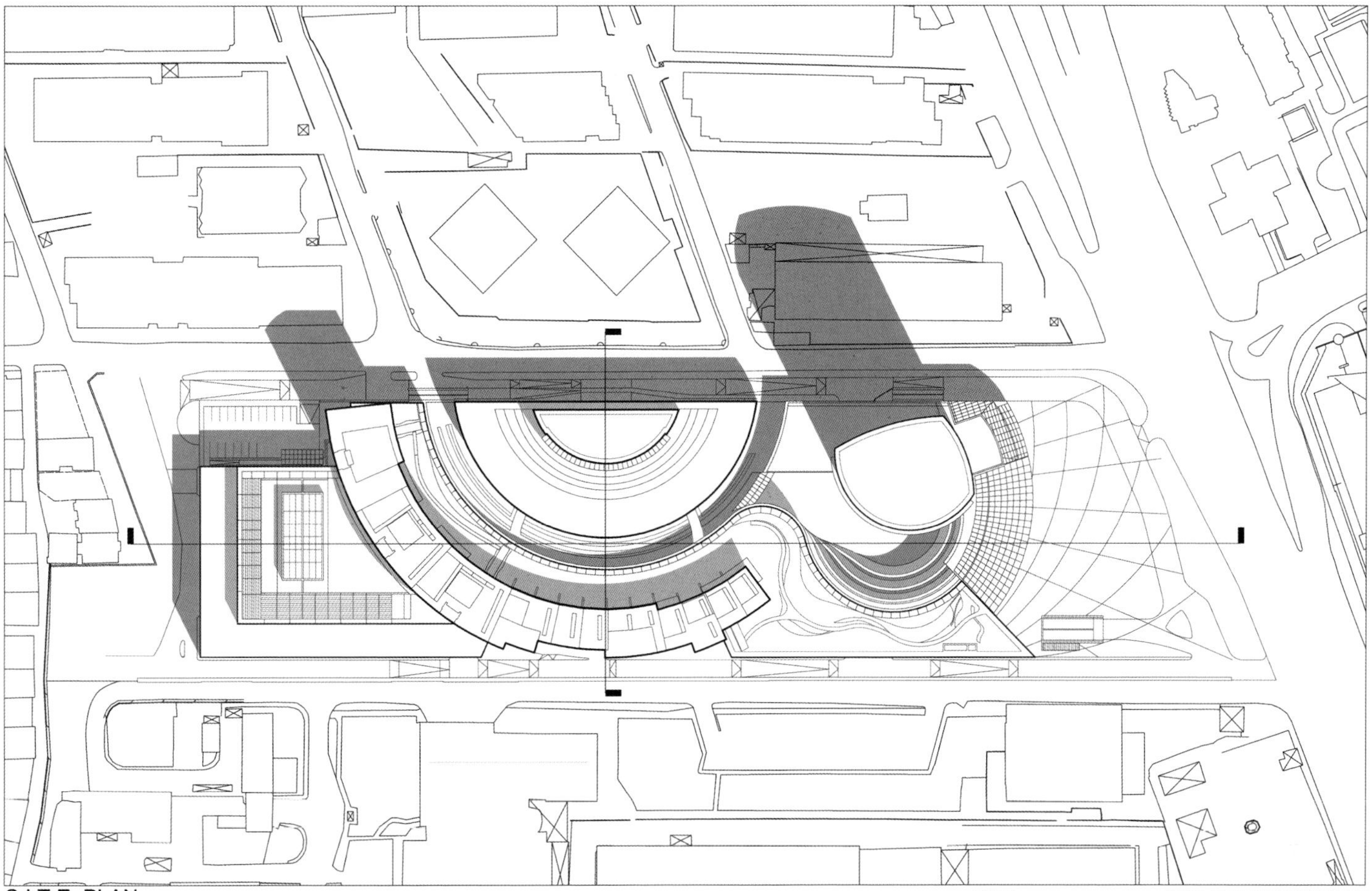

SITE PLAN

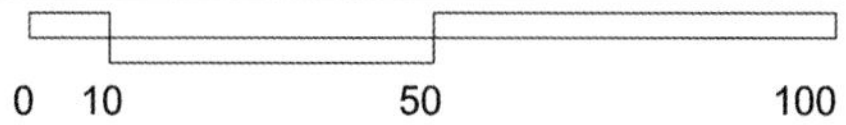

0 10 50 100

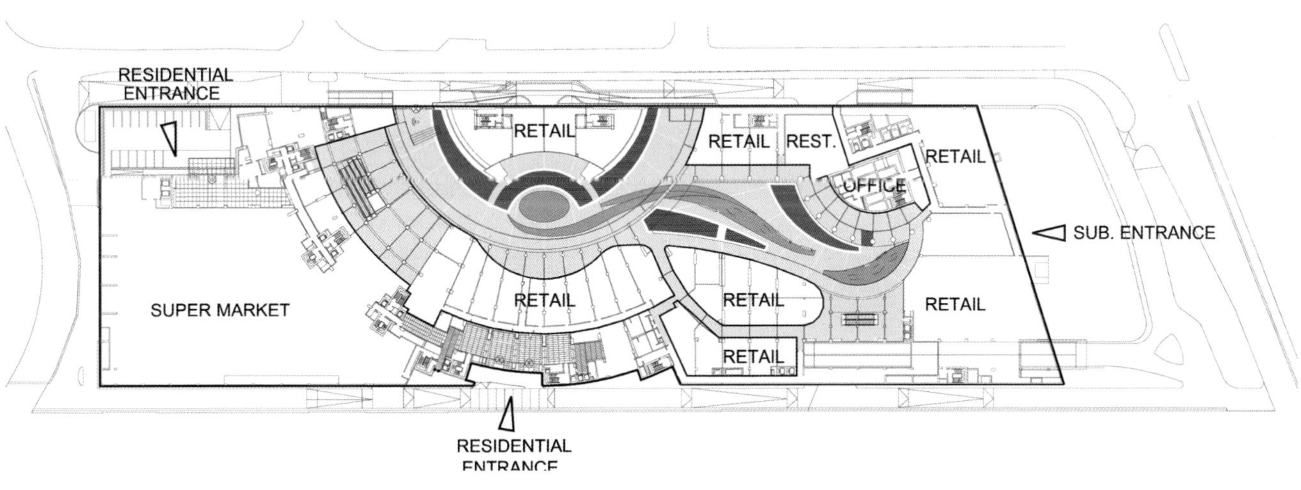


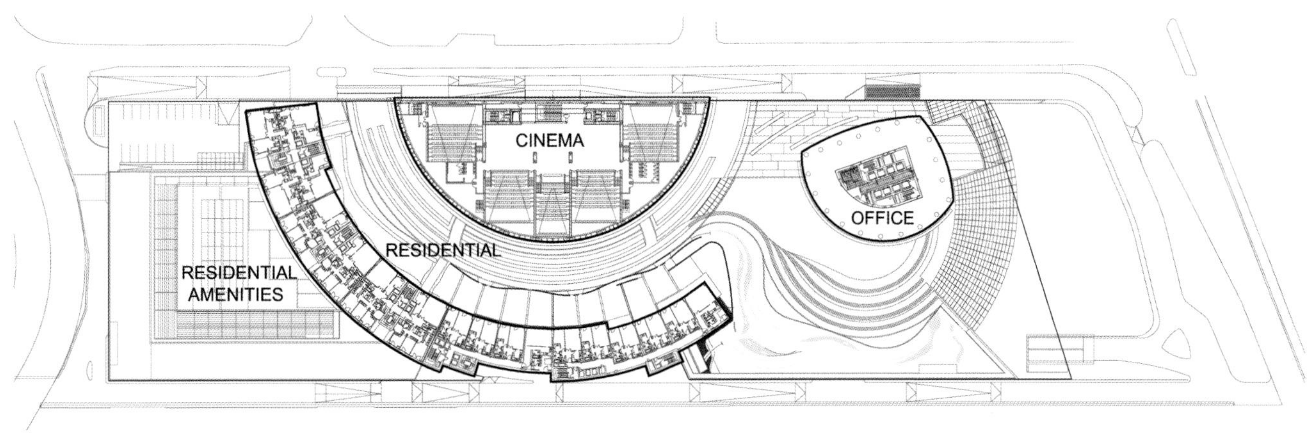

# Las Arenas
# 拉斯阿瑞纳斯

Las Arenas is strategically located at the foothill of Montjuic at the intersection of Gran Via and Avenue Parallel, two major city highways, and acts as a landmark for the Plaça d'Espanya transport interchange. The historic bullring, constructed at the very end of the 19th century, fell largely into disuse towards the end of the 1970s due to the declining popularity of bull fighting in Catalonia. However, the strong civic and cultural role which the building played in the life of Barcelona over nearly a century led to a decision by the city council that the façade should not be demolished.

As a result, the redevelopment has retained and refurbished the striking neo-mudéjar façade, while creating an open and accessible entrance to the new building at street level. Within the façade are commercial, entertainment, health and leisure spaces focused around a central event space.

The original bullring was raised above the levels of the surrounding streets with ramps and stairs within the plinth providing access. However, the redevelopment – which involved the excavation of the base of the facade and the insertion of composite arches to support the existing wall and create new spaces for shops and restaurants – re-establishes a new, open public realm around the base of the building providing level access to the complex.

### Location
Barcelona, Spain

### Area
105,816 sq.m.

### Company
Rogers Stirk Harbour + Partners

### Designers
Rogers Stirk Harbour + Partners

### Photographers
David Cardelús, Josep Mª Molinos, Jan Guell, James Leathem, Eamonn O'Mahony

The telecommunications tower facing Plaça d'Espanya reinforces the presence of the bullring and – at its base - provides direct access from street level and the metro station Espanya to the building. A separate building - the "Eforum" - in Carrer de Llança, provides retail and restaurants at ground and first-floor levels, with four levels of offices above.

The most spectacular aspect of the intervention is the inclusion of a 100 - metre - diameter habitable "dish", floating over the facade of the bullring - and structurally independent from it - providing flexible, column - free spaces beneath. This "plaza in the sky" incorporates large terraces with 360 - degree stunning views over the city around the perimeter of its 76 - metre - diameter domed roof, which covers spaces for cafés and restaurants and a central multi - purpose space. Las Arenas was opened to the public in March 2011.

拉斯阿瑞纳斯地理位置优越，位于蒙特伊克山丘脚下两大主要城市公路格兰维亚大道与Avenue Parallel的交叉处，并充当西班牙广场运输交汇处的路标。大部分建于19世纪末的历史悠久的斗牛场因加泰罗尼亚斗牛的欢迎度下降，在20世纪70年代末被荒废了。然而，因为这座建筑在巴塞罗那超过一个世纪的生活中起着强烈的公民意识和文化的作用，所以市议会决定不拆除门面。

其结果是，重建工程保留并翻新了这引人注目的新穆迪哈尔门面，同时在街道层面上为这栋新建筑建造了一个开放的、便于进入的入口。外墙内包括商业、娱乐、保健和休闲空间，并着重围绕着一个中央活动空间。

原来的斗牛场以坡道和阶梯来提升至高出周围街道层面，在底座之间提供入口。然而，重建工程涉及到门面基地的挖掘和复合拱桥的嵌入以支撑现有的墙并为商铺和餐厅建造新的空间，因此在建筑的基地周围重新建造了一个新型的、开放的公共空间，为建筑提供平坦的入口。

面朝西班牙广场的电信塔强化了斗牛场的存在，在基地上从街道层和西班牙广场地铁站到建筑提供直接的入口。一座位于Carrer de Llan-a的独立建筑“Eforum”的地下层和第一层带有零售和餐厅，楼上4层为办公区。

工程中最为壮观的方面是要建造一个直径为100m并且可居住的“碟”，悬浮于斗牛场门面的上方但结构上独立于它，下方带有极具灵活性的无柱空间。这个“空中广场”带有可360度观看壮丽的城市景色的大露台，围绕着直径为76m的圆顶外围，并包括咖啡厅、餐厅和中央多功能空间。拉斯阿瑞纳斯在2011年3月向公众开放。

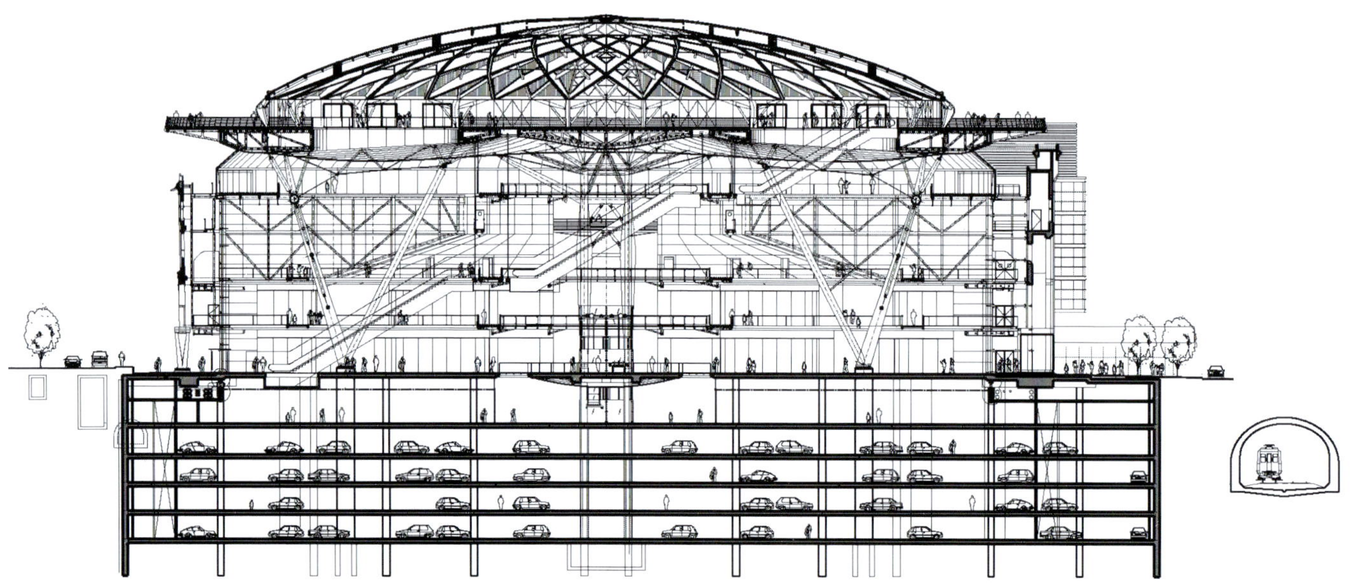

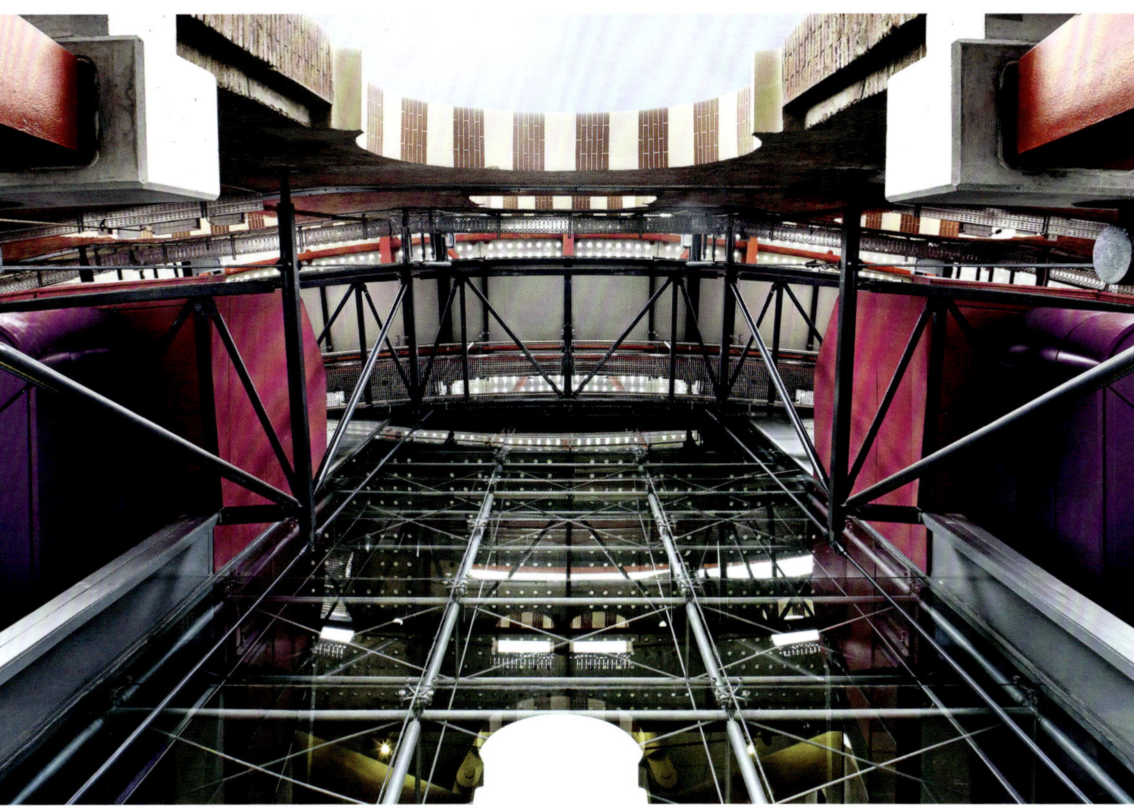

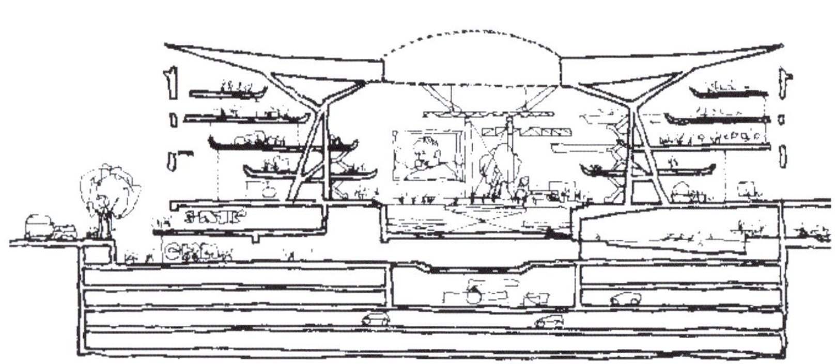

Option A - Building without the use of the roof deck

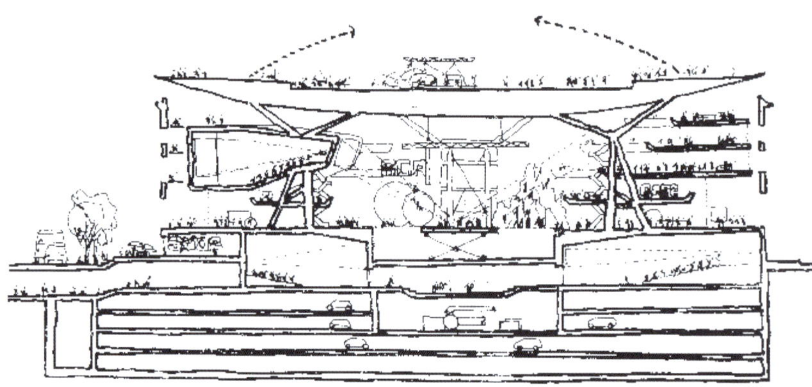

Option B - Building with the use of the roof deck

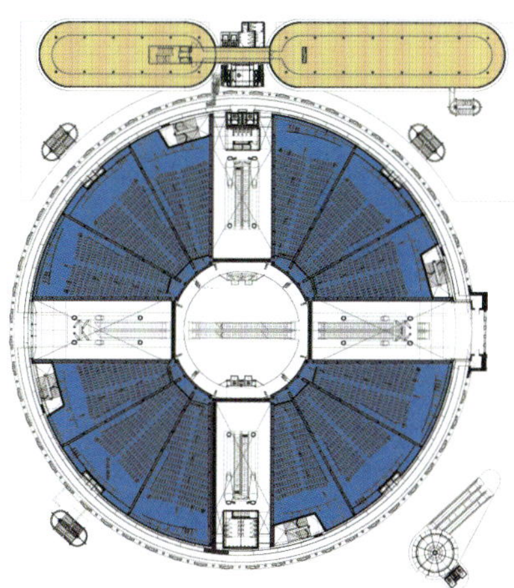

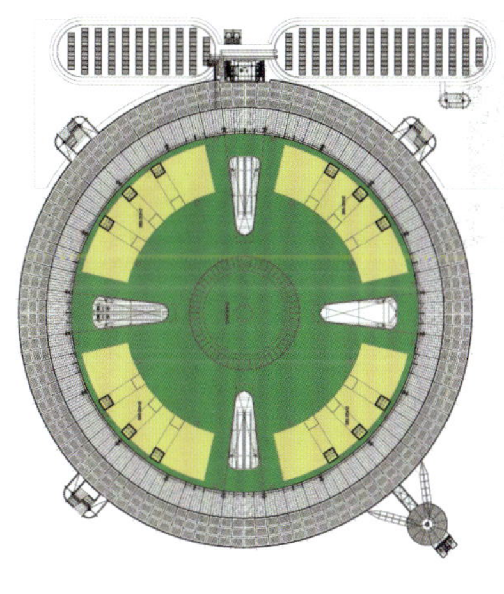
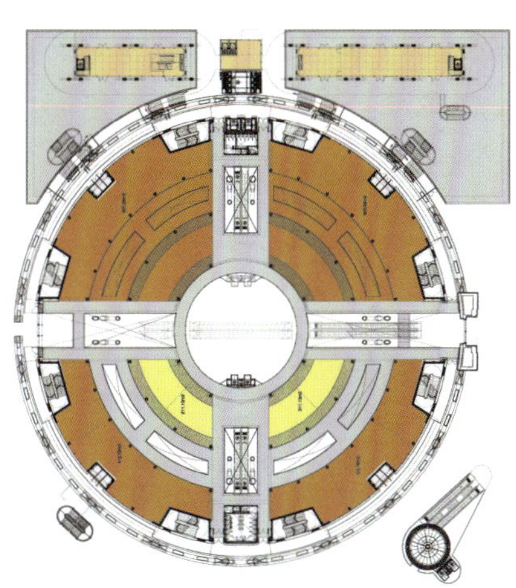
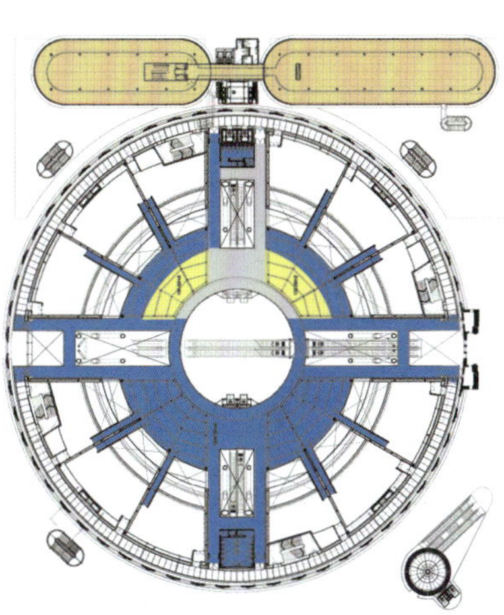
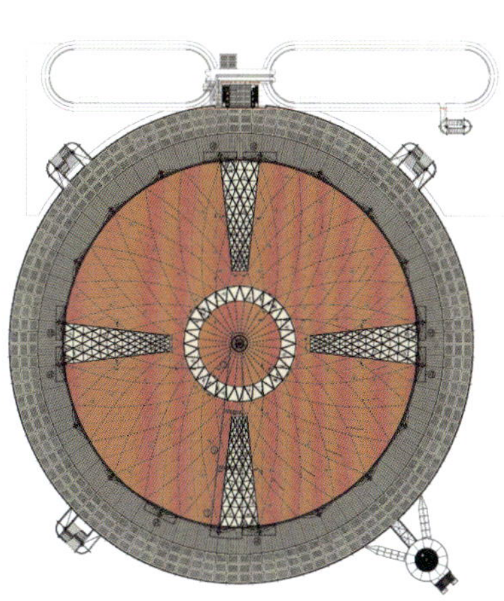

NESPRESSO

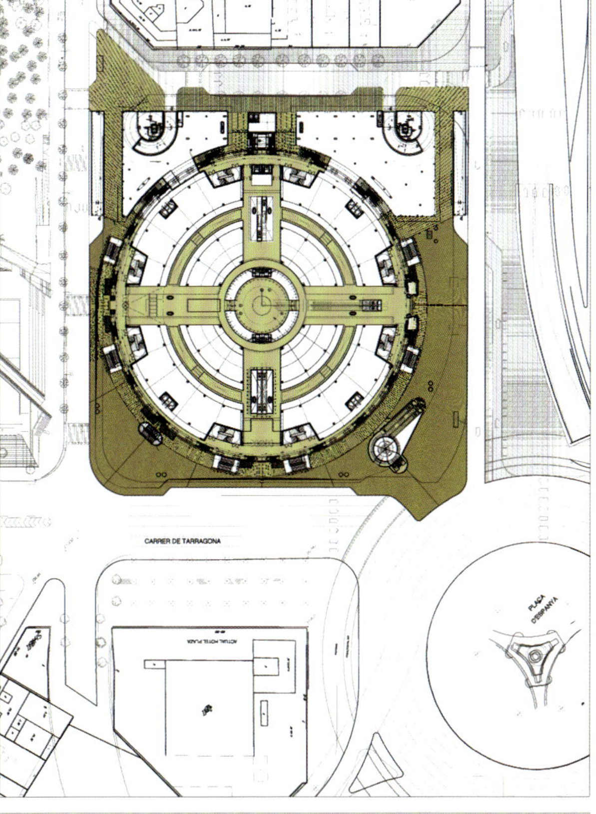
CARRER DE TARRAGONA

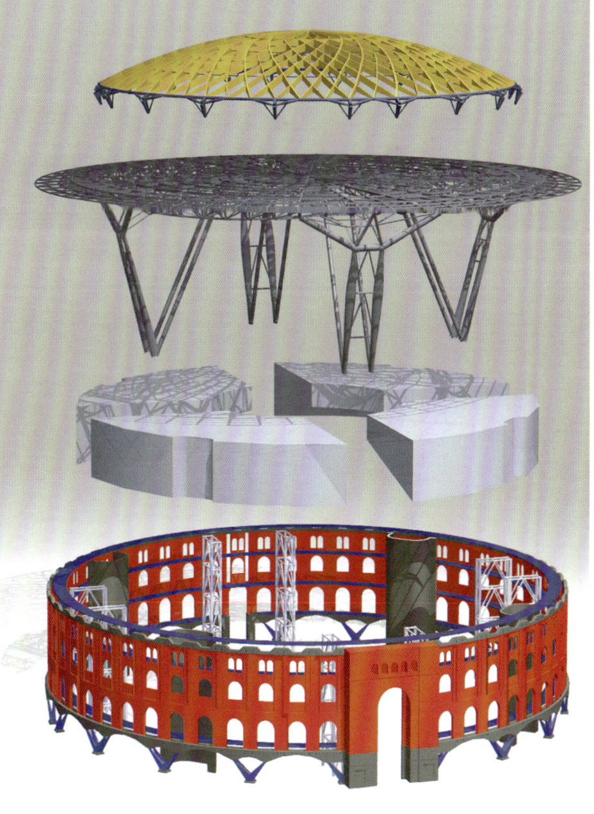

# City Square Mall

# 城市广场购物中心

City Square Urban Park fronts City Square Mall's entrance and provides a recreational environment for nearby residents and visitors alike. In addition, the natural space serves as a constant reminder of the importance of environmental conservation.

The sunken plaza is the focal point of the park, serving as a multi - purpose space while also acting as a passageway linking the MRT station to the mall. An eco - roof - comprising of solar panels, low - E glass panels and a green roof-traps naturally produced energy that in turn powers the lightings, regulates temperature and controls wind circulation. Eco-friendly materials, such as Eco - tiles and recycled timber, were used to construct the park's various structures, such as the playground.

**Location**
Singapore

**Area**
65,642 sq.m.

**Company**
Ong&Ong Pte Ltd

**Designer**
Steven Low

**Photographer**
See Chee Keong, Kajima Overseas Asia

城市公园广场面向城市广场购物中心的入口，并为附近的居民和游客提供了一个休闲环境。此外，该自然空间还作为一个提醒人们保护环境重要性的永久提醒物。

下沉的广场是公园的焦点，既作为一个多用途的空间，同时也作为一个连接地铁站和商场的通道。由太阳能电池板、低辐射玻璃面板和绿色屋顶组成的一个生态屋顶吸收了自然产生的能量，这些能量反过来为灯光提供了动力，调节了温度并且控制了气流循环。环保材料，如生态砖和再生木材，都用来建造公园的各种结构，如操场。

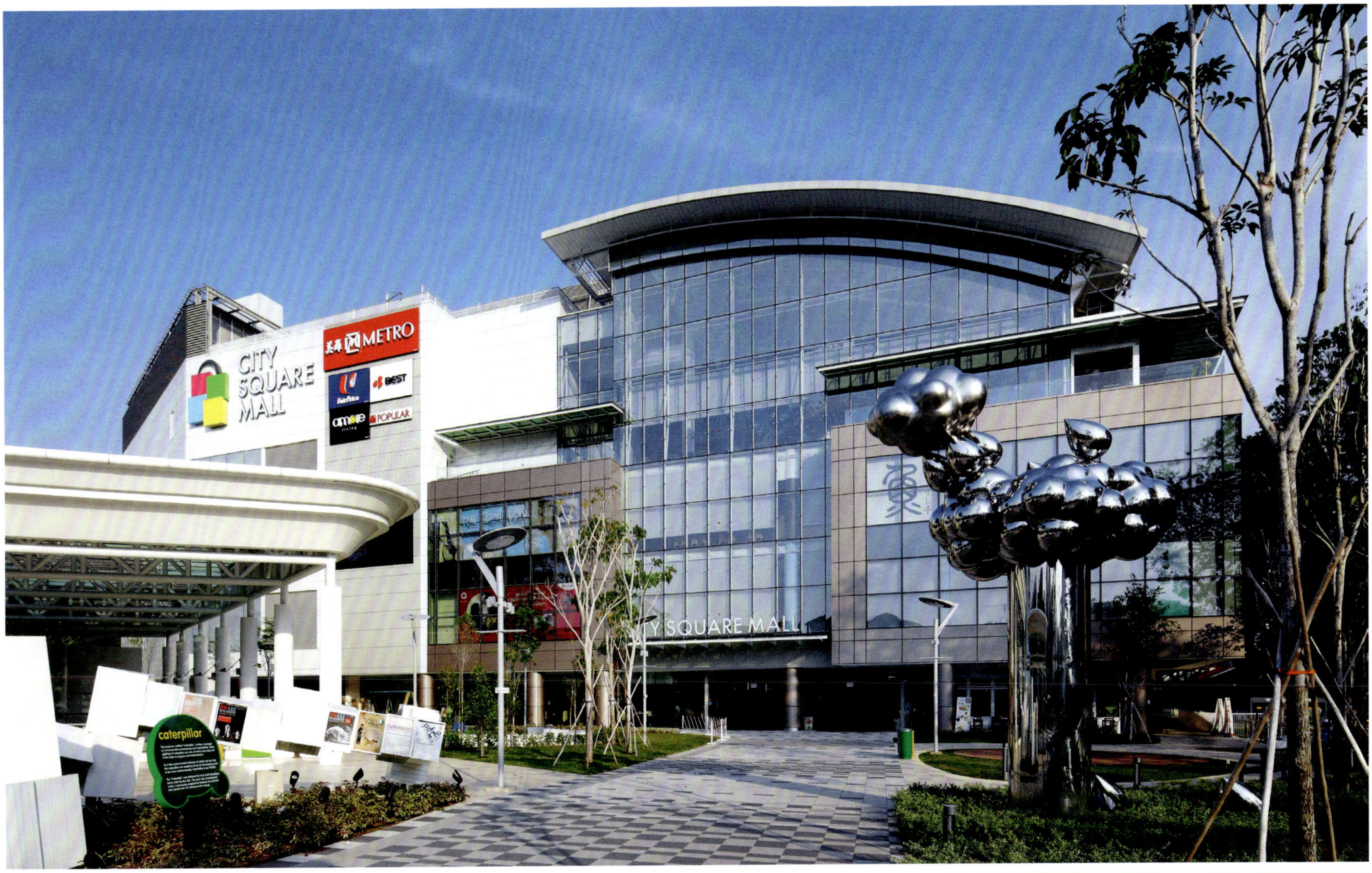
CITY
SQUARE
MALL
METRO
BEST
POPULAR
caterpillar
SQUARE MALL

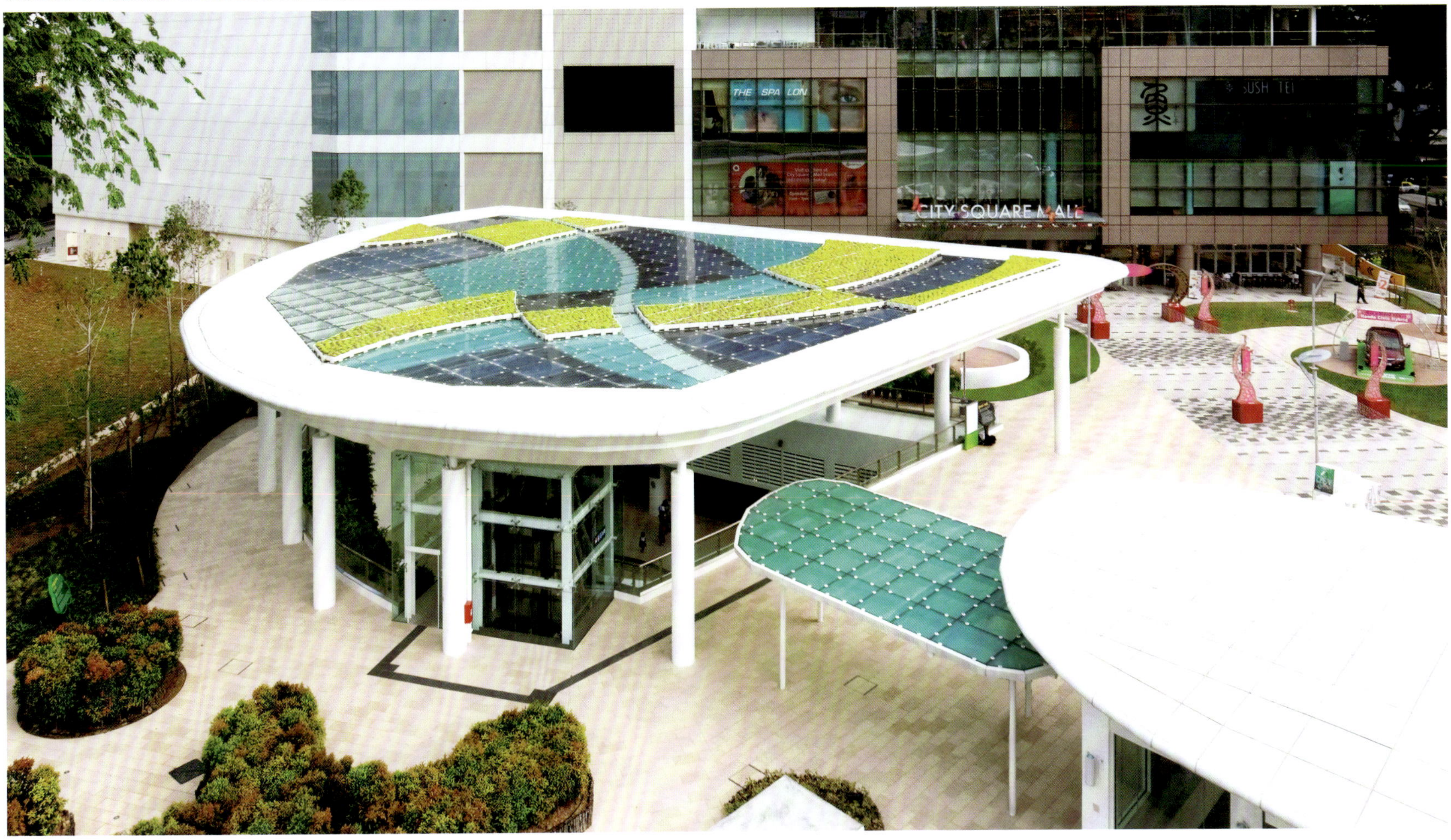
CITY SQUARE MALL

恭賀
新禧
Driveway sensors at B4 carpark help reduce 26 tonnes of $CO_2$ emissions annually.
L3
GIORDANO
PICKET RAIL

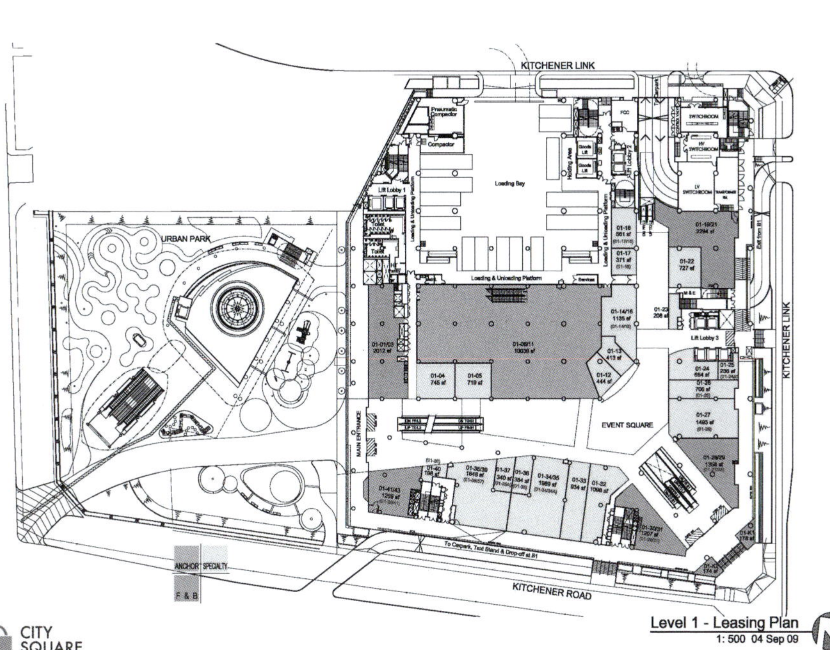
KITCHENER LINK
Pneumatic Compactor
Compactor
Loading Bay
Lift Lobby 1
Loading & Unloading Platform
URBAN PARK
SWITCHROOM
HV SWITCHROOM
LV SWITCHROOM
01-19/21 2294 sf
01-22 727 sf
01-14/16 1135 sf
01-01/03 2012 sf
01-08/11 13036 sf
01-04 745 sf
01-05 719 sf
01-12 444 sf
01-13 413 sf
01-24 664 sf
01-27 1493 sf
Lift Lobby 3
EVENT SQUARE
MAIN ENTRANCE
KITCHENER LINK
01-41/43 1259 sf
01-34/35 1989 sf
01-33 934 sf
01-32 1098 sf
01-30/31 1207 sf
01-28/29 1358 sf
To Carpark, Taxi Stand & Drop-off at B1
KITCHENER ROAD
ANCHOR
SPECIALTY
F & B
CITY SQUARE MALL
Level 1 - Leasing Plan
1: 500 04 Sep 09
1. Indicative plans for reference only, subject to further changes and authorities' compliances.
2. This drawing to be read in conjunction with the Builder's Construction/shop drawings/ Architects Design Intent/Submission & Engineers' drawings.

OVER 1,000
NOW OPEN

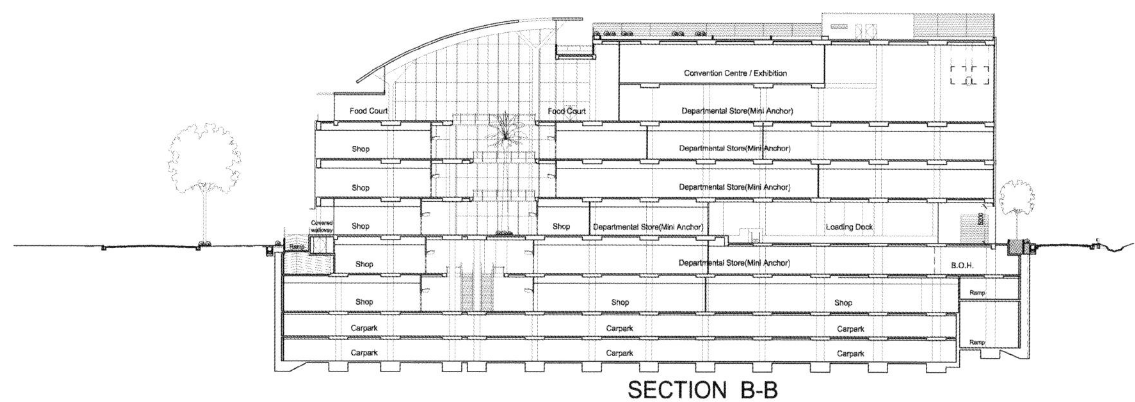
Convention Centre / Exhibition
Food Court
Food Court
Departmental Store(Mini Anchor)
Shop
Departmental Store(Mini Anchor)
Shop
Departmental Store(Mini Anchor)
Shop
Shop
Departmental Store(Mini Anchor)
Loading Dock
Shop
Departmental Store(Mini Anchor)
B.O.H.
Shop
Shop
Shop
Carpark
Carpark
Carpark
Carpark
Carpark
Carpark
Ramp
SECTION B-B

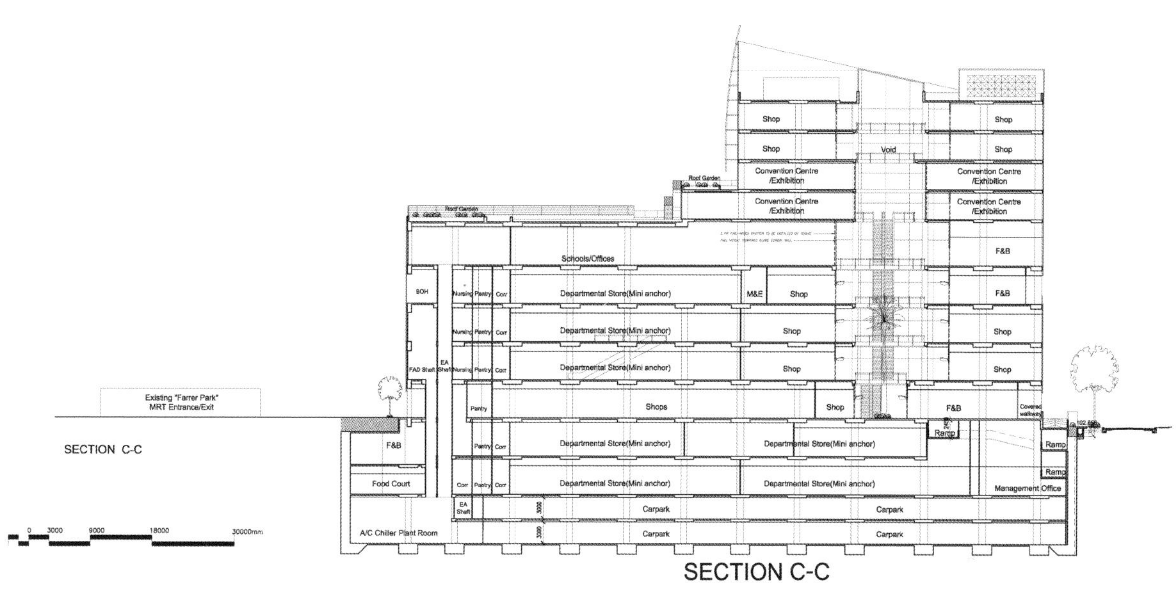
Shop
Shop
Shop
Shop
Void
Convention Centre /Exhibition
Convention Centre /Exhibition
Convention Centre /Exhibition
Convention Centre /Exhibition
Roof Garden
Schools/Offices
F&B
Departmental Store(Mini anchor)
M&E
Shop
F&B
Departmental Store(Mini anchor)
Shop
Shop
Departmental Store(Mini anchor)
Shop
Shop
Shops
Shop
F&B
Existing "Farrer Park" MRT Entrance/Exit
SECTION C-C
F&B
Departmental Store(Mini anchor)
Departmental Store(Mini anchor)
Ramp
Food Court
Departmental Store(Mini anchor)
Departmental Store(Mini anchor)
Management Office
Carpark
Carpark
A/C Chiller Plant Room
Carpark
Carpark
0 3000 9000 18000 30000mm
SECTION C-C

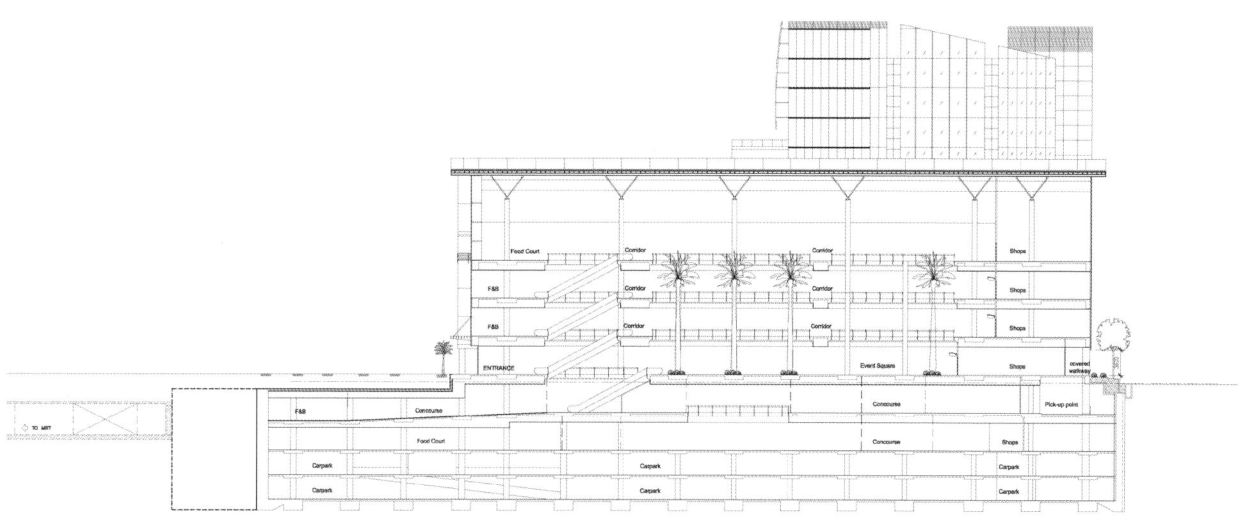
Food Court
Corridor
Corridor
Shops
F&B
Corridor
Corridor
Shops
F&B
Corridor
Corridor
Shops
ENTRANCE
Event Square
Shops
F&B
Concourse
Concourse
Pick-up point
Food Court
Concourse
Shops
Carpark
Carpark
Carpark
Carpark
Carpark
Carpark

CyberActive
CITY
help

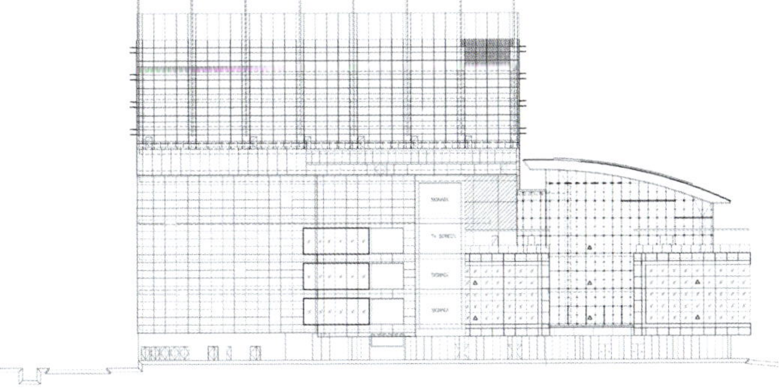

NORTH WEST ELEVATION
Elevation from Serangoon Road

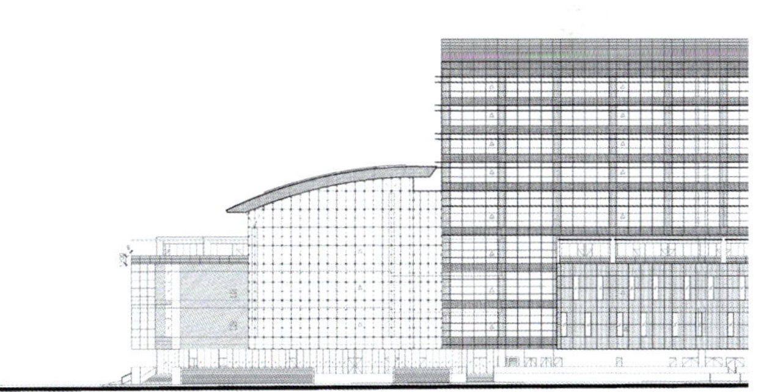

SOUTH EAST ELEVATION
Elevation from City Square Residences

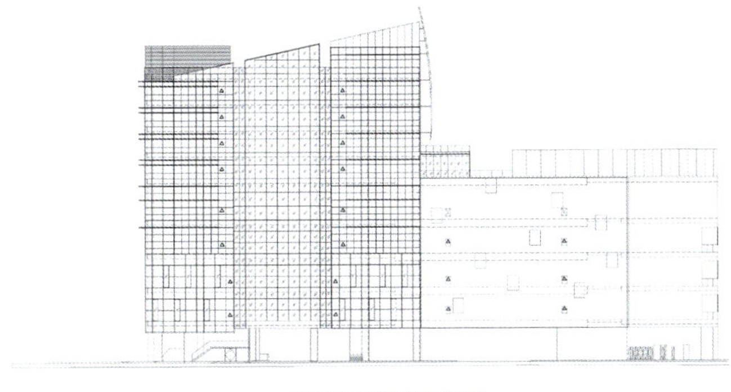

NORTH EAST ELEVATION
Elevation from service road

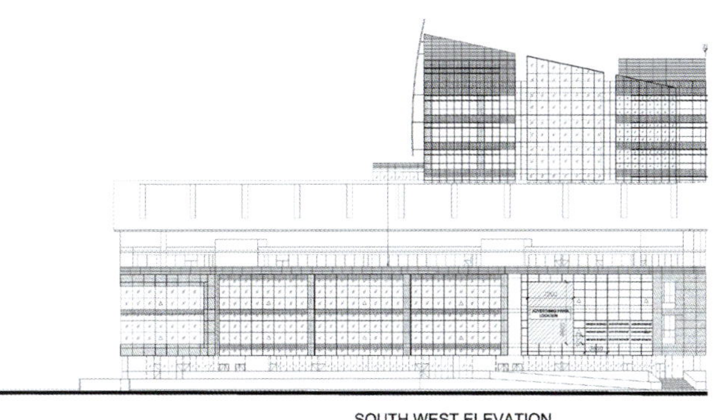

SOUTH WEST ELEVATION
Elevation along Kitchener Road

# ION Orchard

## ION Orchard购物中心

Benoy's design for ION Orchard, we sought inspiration from the site's historical function as an orchard. Specifically, Benoy focused on the metaphor of gestation - a process that is at the heart of the natural environment. The idea of growth is central to the nature of an orchard. Intrinsic to this organic process is the idea of peeling, which suggests the emergence of something new; the removal of an external skin to reveal growth that has taken place within.

The plan of this mixed-use scheme makes direct reference to the contours and lines of fruit and nuts. Meanwhile, the retail canopy draws on the metaphors of "skin", "canopies of trees", "foliage" and "fruit and peels", and the iconic tower design is based on the concept of "roots and shoots"- to celebrate the site, its history and its future.

By linking all visual elements to the site - drawing on the imagery of the natural environment and of Singapore itself, Benoy has tied the development to the place. By including in the scheme grand civic gestures and new, flexible public spaces (including art gallery and observation deck), it has been brought closer to the people.

### Location
At the gateway to Orchard Road, at the junction of Orchard Road and Paterson Road

### Area
61,316 sq.m.

### Company
Benoy Architect

### Designer
Benoy, David Buffonge, Sarah Lee, Axel Funke, Jason Hall

在ION Orchard的设计中，Benoy 设计公司从该旧址原是一座果园的历史功能中获得了灵感。具体说来，Beony设计公司专注于将建筑过程比喻成妊娠——这一过程是自然环境的中心。生长的概念是一座果园性质的核心。这一有机过程的内在本质就是“剥离”的思想，这意味着新事物的出现；去掉外皮揭示出已经发生的内部增长。

这种混合使用计划的平面图直接引用了水果和坚果的轮廓与线条。同时，商店的顶棚援引了“皮肤”、“树的华盖”、“树叶”以及“水果和果皮”的比喻，具有标志意义的塔形设计则是基于“根与芽”的概念——以颂扬该建筑所在的地址，它的历史和未来。

通过将各种视觉因素链接到该地址，借鉴自然环境和新加坡本身的意象，Benoy设计公司将发展紧紧系于该地区。在计划中纳入宏伟的公民姿态和新的、灵活的公共场所（包括艺术画廊和景观台），使得商场更贴近人民。

ION
ORCHARD

Cartier
Cartier
Cartier
Cartier
Cartier
ion
ORCHARD

Cartier
Cartier
Cartier

L'OCCITANE

ONE THREAD MAKES A STITCH.
A WEAVE OF THREADS MAKES A NEW YOU.
WHERE

IWC
BALLY
Z Zegna

# Zeilgalerie Frankfurt - Façade

## 法兰克福市采尔购物长廊外观

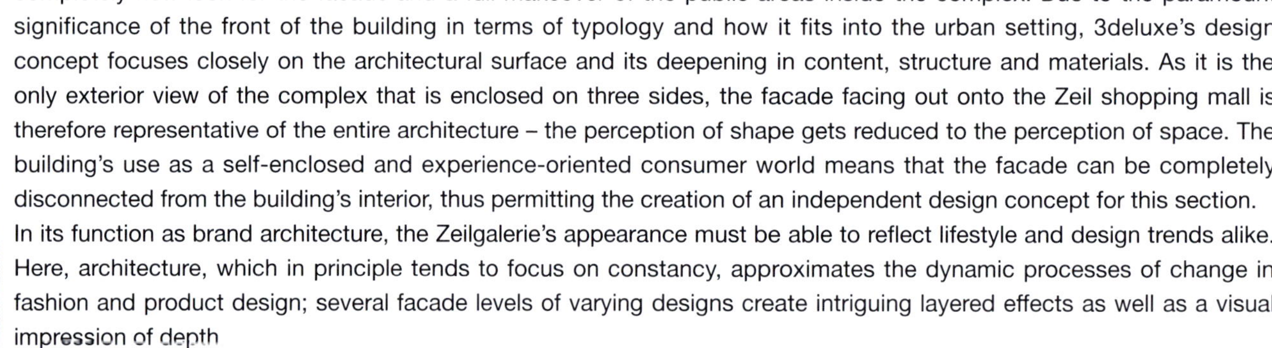

The redesign of the Zeilgalerie shopping centre, which is renowned far beyond Frankfurt's city limits, includes a completely new look for the facade and a full makeover of the public areas inside the complex. Due to the paramount significance of the front of the building in terms of typology and how it fits into the urban setting, 3deluxe's design concept focuses closely on the architectural surface and its deepening in content, structure and materials. As it is the only exterior view of the complex that is enclosed on three sides, the facade facing out onto the Zeil shopping mall is therefore representative of the entire architecture – the perception of shape gets reduced to the perception of space. The building's use as a self-enclosed and experience-oriented consumer world means that the facade can be completely disconnected from the building's interior, thus permitting the creation of an independent design concept for this section.

In its function as brand architecture, the Zeilgalerie's appearance must be able to reflect lifestyle and design trends alike. Here, architecture, which in principle tends to focus on constancy, approximates the dynamic processes of change in fashion and product design; several facade levels of varying designs create intriguing layered effects as well as a visual impression of depth

**Location**
Frankfurt/Main, Germany

**Main Materials**
Aluminium sheets, powder coated, Laminated glass panes, Steel, etc.

**Company**
3deluxe transdisciplinary design

**Designers**
3deluxe transdisciplinary design

**Photographer**
Emanuel Raab

采尔购物长廊，其知名度远远超出法兰克福的城市边界，它的重新设计包括门面的全新样貌和综合楼里的公共区域的全面改造。由于楼前在类型与它如何适应城市环境方面有着重要意义，设计师的设计理念紧紧围绕着建筑表面和它在内容、结构和材料方面的深化。由于它是三面封闭的综合楼的唯一外观，因此采尔购物长廊朝外的外观代表着整个建筑——形状认知降低到空间认知。作为一个自我封闭和经验为主的消费世界，建筑的使用意味着门面可从建筑内部完全分离，从而建立了一个独立的设计理念。

在其作为品牌架构功能的同时，采尔购物长廊的外观必须能体现生活方式以及设计趋势。在这里，原则上侧重于恒久不变的建筑，接近于时装以及产品设计中变化的动态过程；不同设计的几个门面层次创造了有趣的分层效果以及深度的视觉印象。

ZEILGALERIE

ZEILGALERIE

Zeil GALERIE

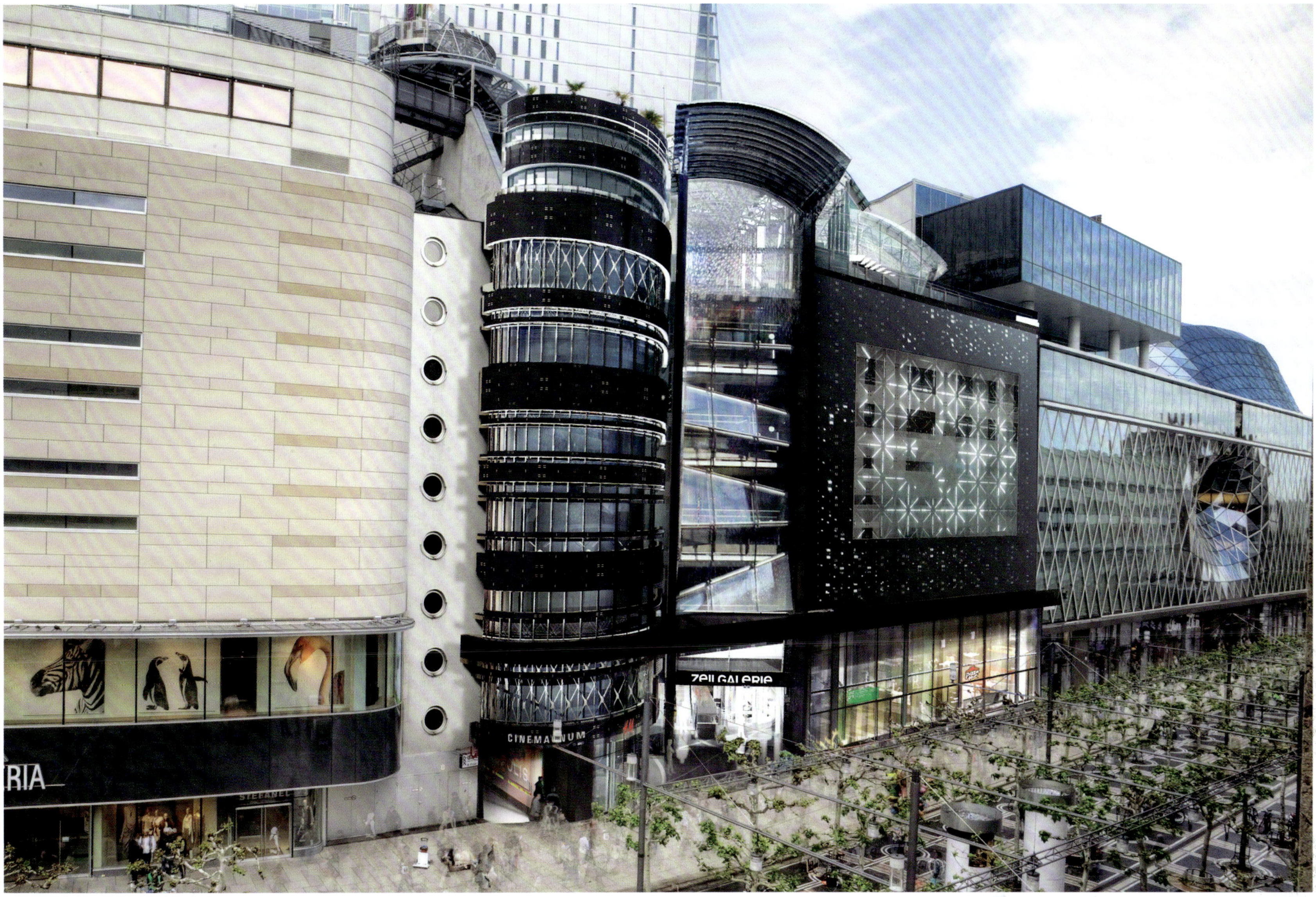

zeil GALerie
WILLKOMMEN

RESTAURANT

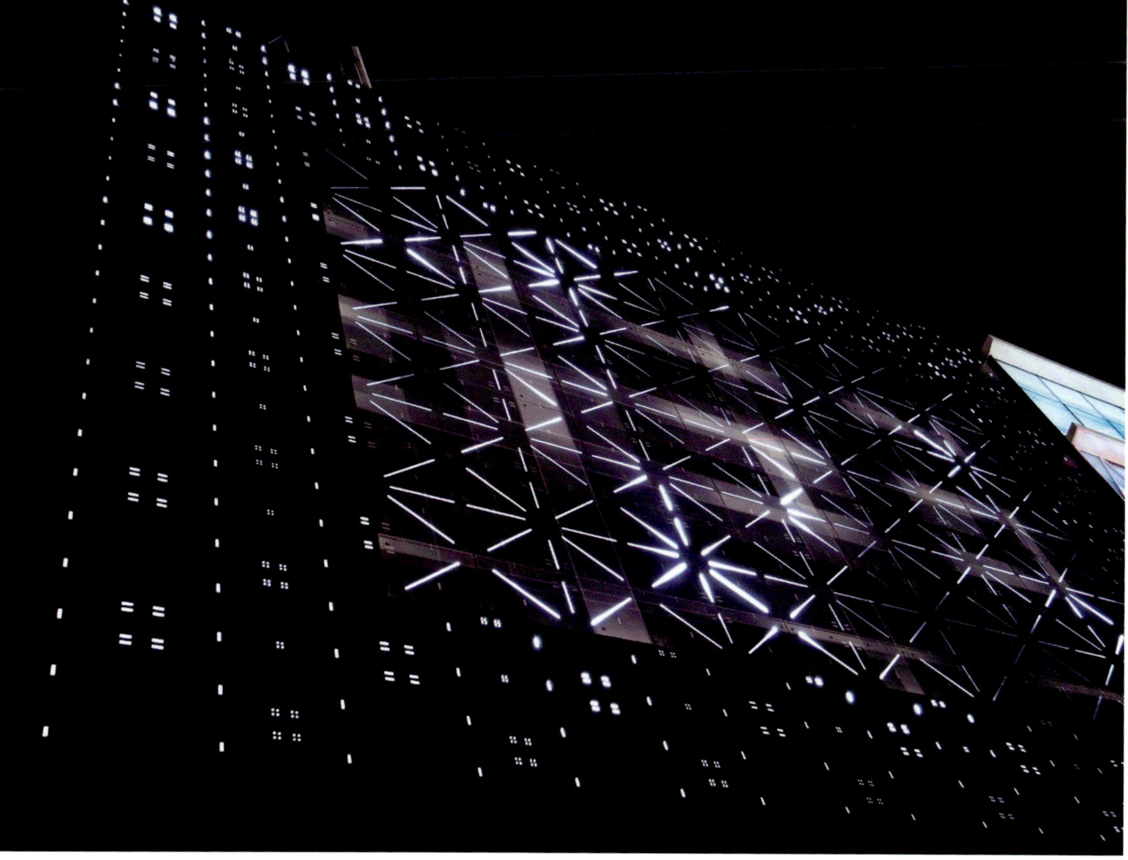

# Elements Mall
# 元素购物中心

Benoy has kept a focus on a design that delivers a holistic experience while simultaneously creating surprising individual experiences. Telling a story through architecture, the design shapes the journey through the space. Changing colors, shapes and volumes throughout Elements constantly invite visitors to explore and discover their surroundings. 40,000 sq ft of skylights will give the interior more natural daylight than any other mall in Hong Kong. The skylights connect nature to the retail experience - offering natural light in as well as a visual link to the rooftop park. Meanwhile, inside the mall, shoppers can engage in a complete social and leisure experience - shopping at high - end retail stores, skating on ice, then wining and dining alfresco - style at the rooftop restaurants. Large public events spaces for live local and international performances, shows and exhibitions have also been incorporated,ensuring cultural vitality is continuously injected into the space.

Asian and Western mall consumers are generally well-traveled and sophisticated in their tastes so they are looking for more than simple retail: they require good restaurants, alfresco dining, hotel quality service and rest rooms, bright and interesting parking areas, arts and culture, engaging design and graphics. In short, they are demanding more"experience" than ever before from their shopping trip, so that it has now evolved to become a "lifestyle day out".

**Location**
Kowloon Station, Hong Kong

**Area**
82,750 sq.m.

**Company**
Benoy Architects

**Designer**
Chris Lohan – Divisional Director

Benoy设计团队一直专注于能传达出整体经验的设计，同时创造出令人惊讶的个人经验。 该设计通过建筑来讲述一个故事，通过空间来塑造出整个旅程。不断变换的色彩、形状以及体积贯穿整个元素购物中心，不断吸引游客探索和发现自己的周围环境。相比香港其他的购物中心，这个40,000平方英尺的天窗将给该室内空间带来更多的自然光。天窗将自然与购物体验联系起来，提供自然光的同时也提供了一个与天台花园的视觉连接。同时，在商场内部，顾客可以拥有一个完整的社交和休闲体验——在高档次的零售商店购物，滑冰，然后在屋顶餐厅露天品酒，就餐。直播本地和国际演出，表演和展览的大型公益活动区也被收纳其中，从而保证文化的生命力不断注入到该空间内。

亚洲和西方商场的消费者一般都游历甚广，自有其高雅的品味，因此他们寻找的绝不是简单的零售店：他们要求有好的餐厅，露天用餐区，酒店似的优质服务和休息室，明亮而又充满情趣的停车场，艺术文化，独特的设计和图形。简而言之，较之以前的购物之旅他们要求得到更多的体验，因此现在购物已经演变成了一种“日常的生活方式”。

ELEMENTS

CANALI
ASCOT CHANG

DIESEL
ANNE KLEIN

NAUTICA

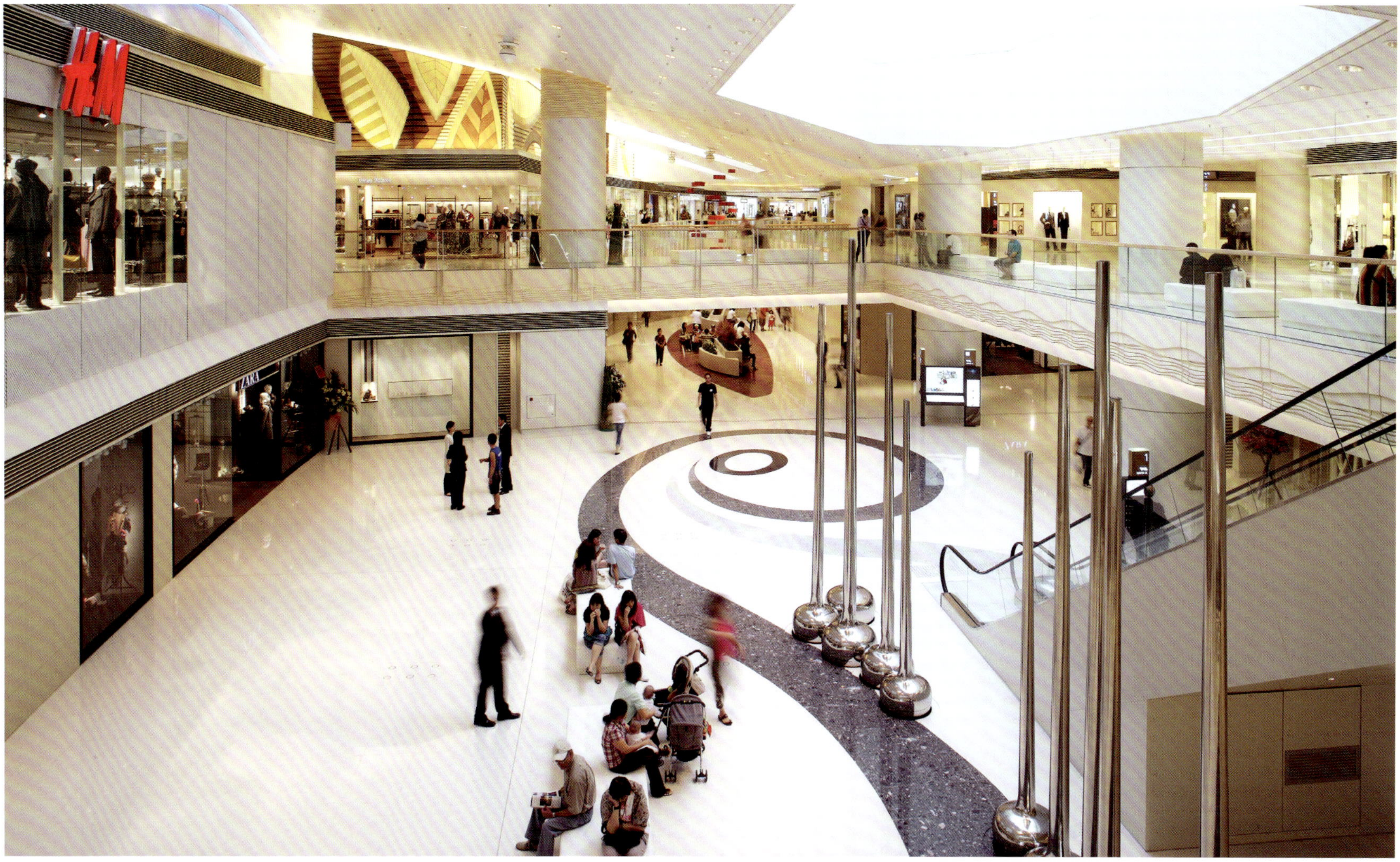

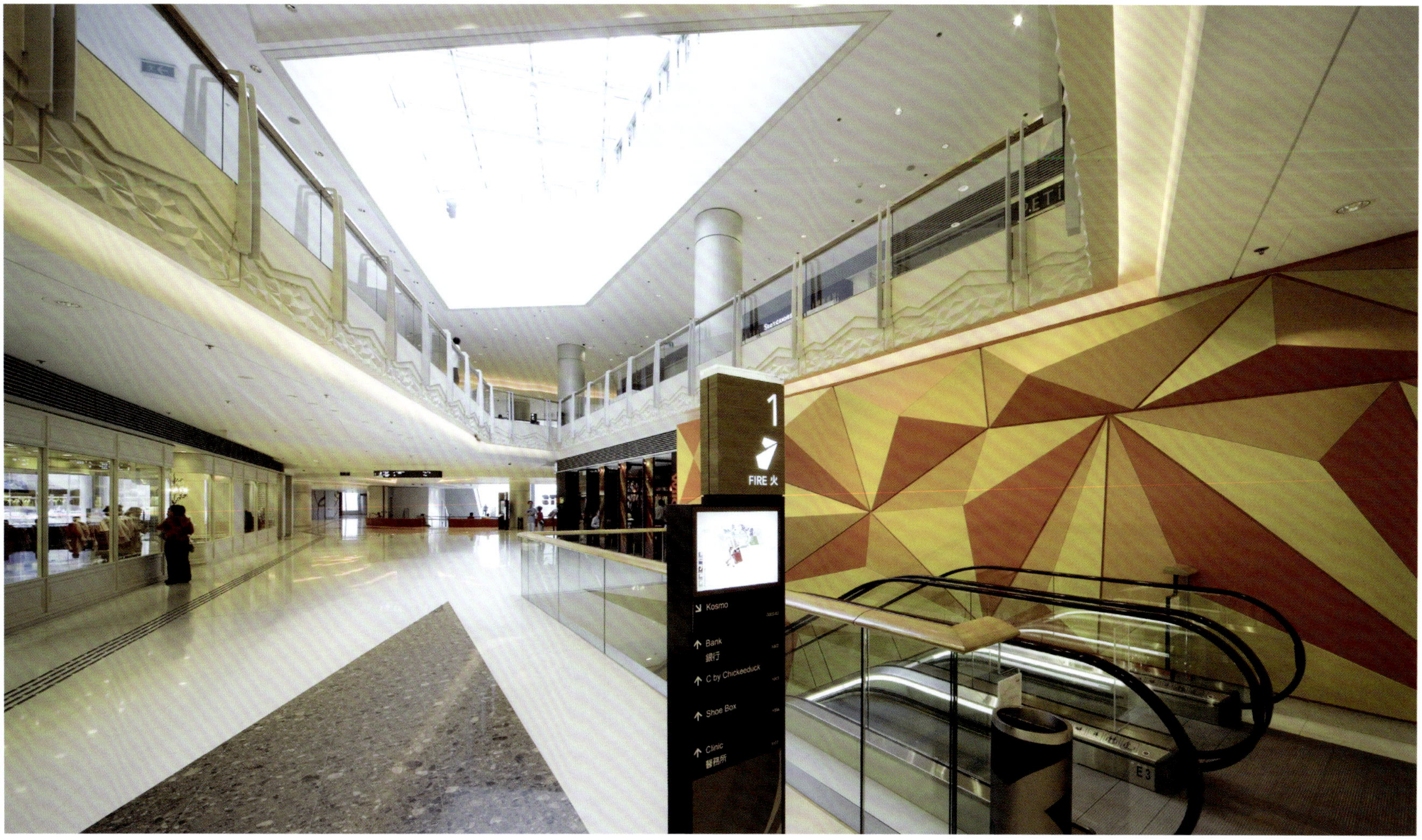

LANVIN
PARIS
TIFFANY&Co.
EXIT

# Ganghui Xin Tian Di

## 港惠新天地

The design scheme of "Huigang Xin Tian Di" Shopping Mall aims to create a new, stylish and modern Shopping Mall with international standards, and the "leisure holiday concept" is applied to the shopping center design.

Through careful analysis of building structure and functional distribution of the mall, in the access node design of north and south two main mall entrances, the designers refer to the point, line and plane of component element of "Hong Kong Wai World" logo VI design. The use of point, line and plane creates shape and color for space. The eye - catching triangular geometry and generous square are used on the wall and ceiling.

From the project's structural analysis, it has various advantages, especially the two huge atrium. So designers use bare, cold, modern slope roof made of steel and glass and metal and glass panel walls to create the perfect casual shopping space. The sun is introducedinto the interior from the top and the light produces different effects at different times. Atrium and the bridges play a role in communication and distributary and add bright colors for the overall environment. At this point, natural light, ecology, leisure, green and health constitute the common theme of Shopping Mall. Original design of the atrium and entrance, delicate bridges and the changing vision, coupled with high - end store design, from indoor to outdoor, all reflect the theme of the Shopping Mall about shopping concept and also reflect the flourishing style in Huizhou.

**Location**

广东惠州

**Area**

4,500 sq.m.

**Main Materials**

彩釉玻璃、彩色钢化玻璃、夹胶钢化清玻、铝单板、外墙砖、灰麻石材等

**Company**

深圳市汉筑装饰设计有限公司

**Designer**

康华

“港惠新天地” Shopping Mall设计方案的主要目的打造一座全新、时尚、现代、具有国际水准的Shopping Mall，并把“休闲假日理念”运用到购物中心设计中。

通过对商场的建筑结构和功能分布的精心分析,在商场的南北两个主要入口节点设计上，设计师参考了“港惠新天地”的VI标识设计的构成元素中的点线面。运用点线面营造空间的造形和色彩，把醒目的三角几何形体和方正大气的正方形用在墙和天花上。

从本项目的建筑结构分析来看，它具有各种优势，特别是两个巨大的中庭。因此设计师以裸露、冷酷、现代感的斜面钢架玻璃屋顶和金属材料玻璃幕墙营造完美的休闲购物空间。把阳光从顶部导入室内，并在不同的时间光线产生不同的效果。中庭和廊桥起到了沟通分流的作用同时也为整体环境增添了亮色。此时，自然光、生态、休闲、绿色、健康构成了Shopping Mall的共同主题。独具匠心的中庭和入口设计，精巧的廊桥，变换的视野，加上高档的店面设计，从室内到室外，无不体现了Shopping Mall的购物概念主题，也体现了惠州的繁华风貌。

港惠购物中心

YINER

Theme
April
幕
果体验店
3151

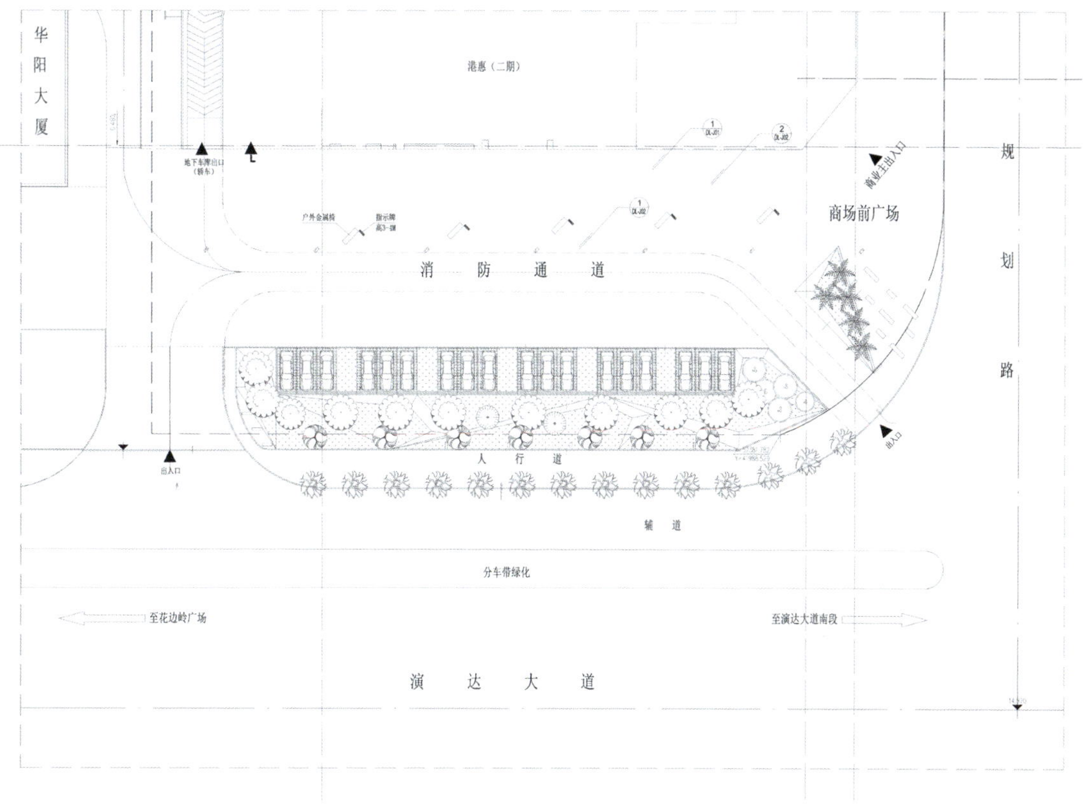

华阳大厦
港惠（二期）
地下车库出口
（轿车）
户外金属椅
指示牌
高3-4M
商业主出入口
商场前广场
消防通道
规划路
出入口
人行道
出入口
辅道
分车带绿化
至花边岭广场
至演达大道南段
演达大道

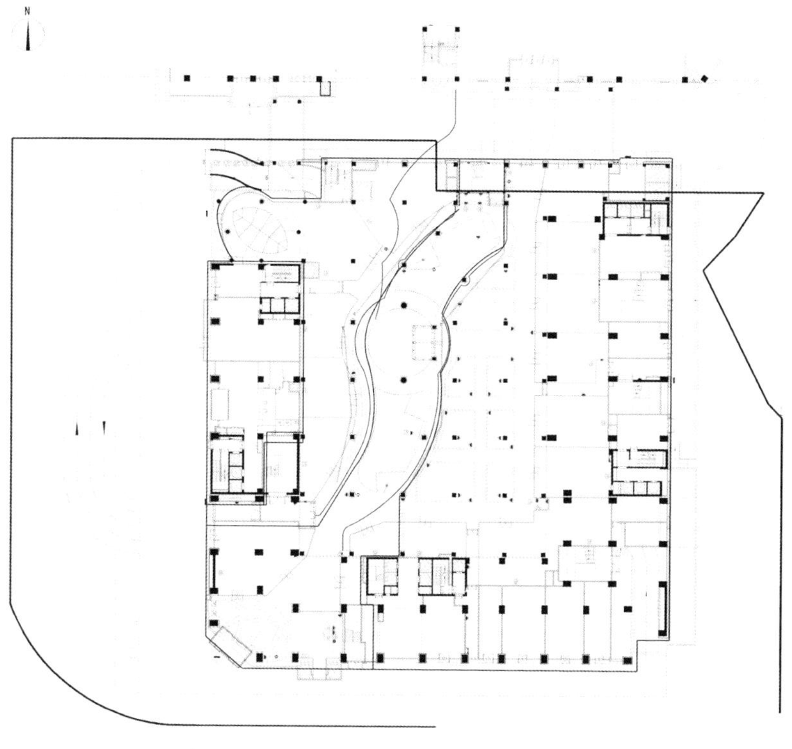

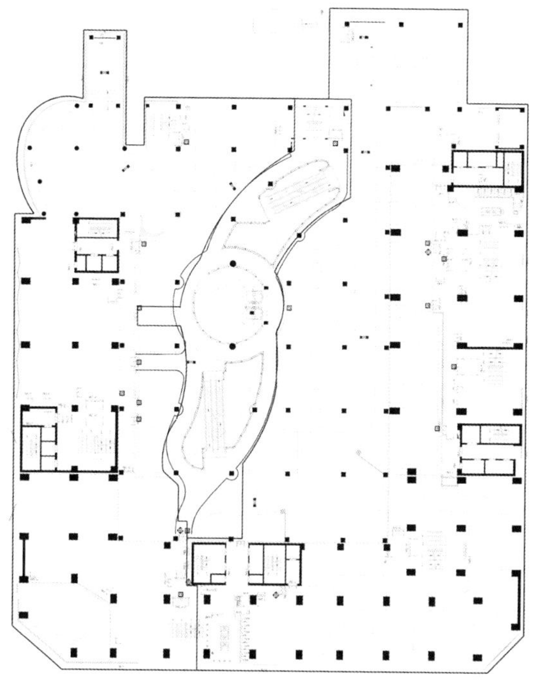

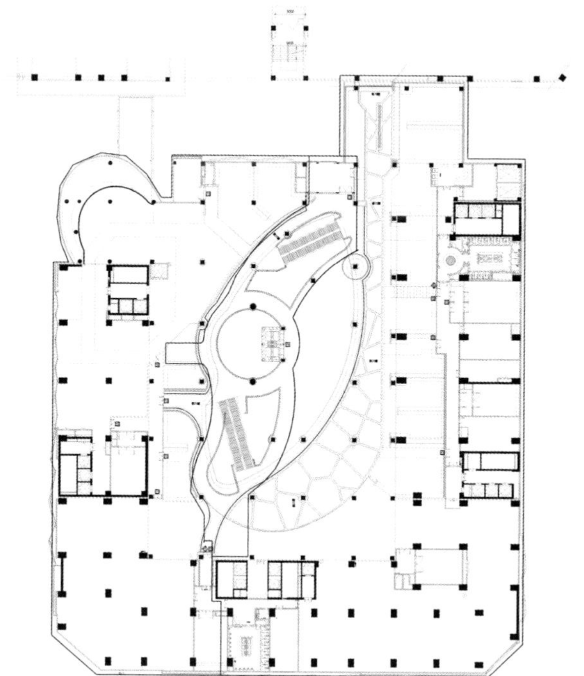

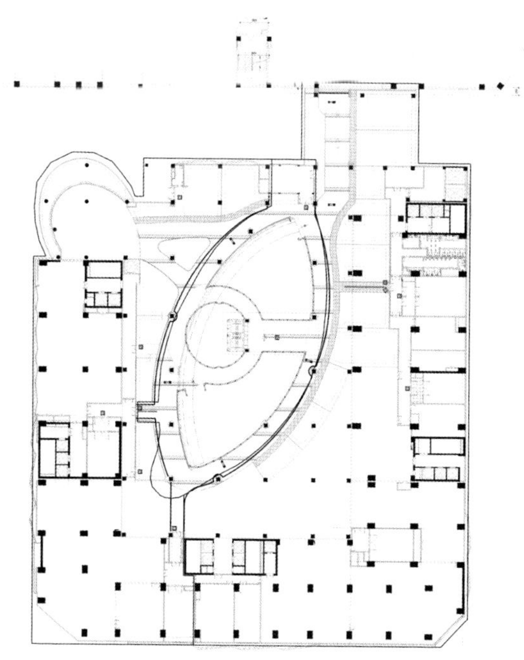

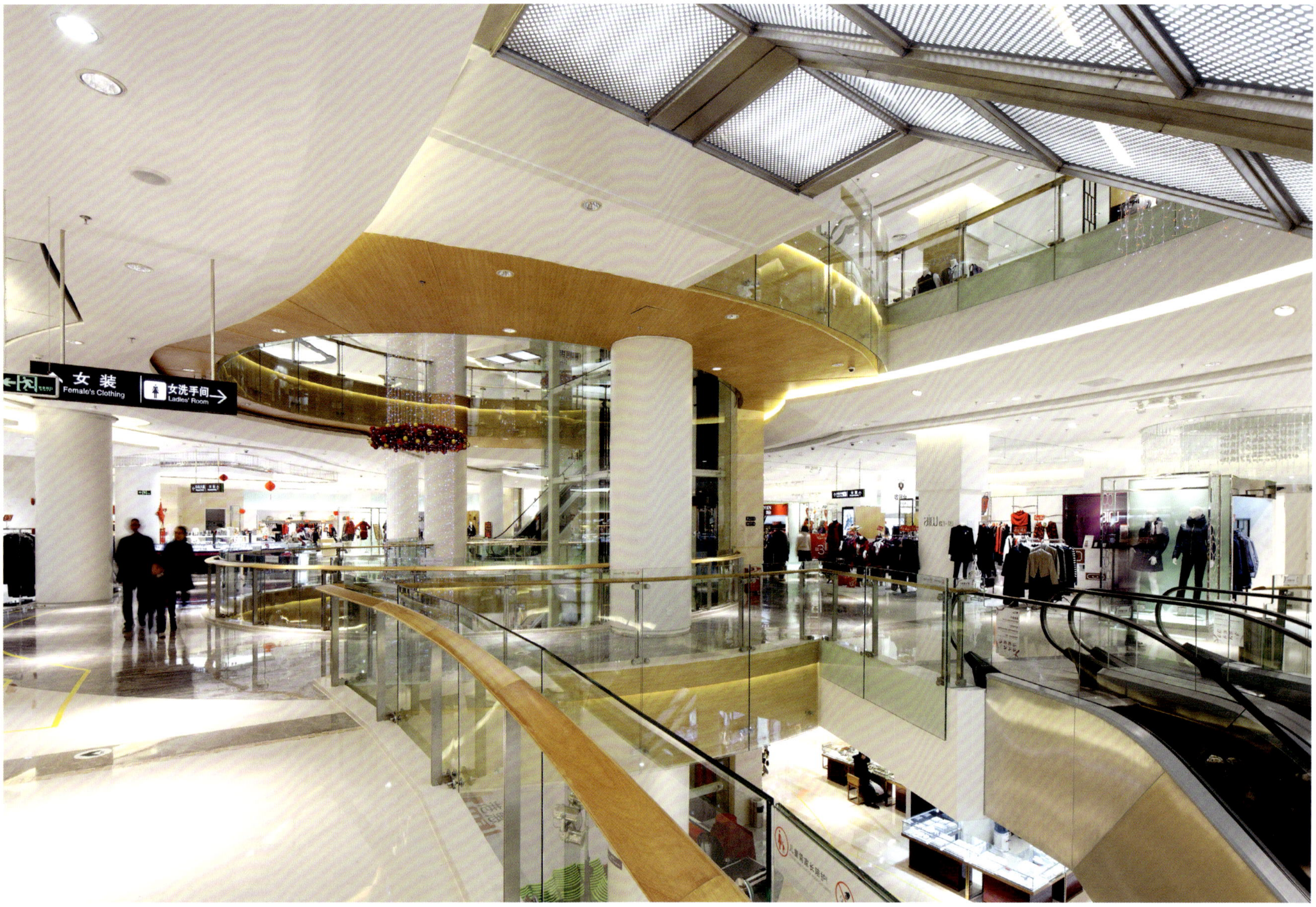
女装
Female's Clothing
女洗手间
Ladies' Room

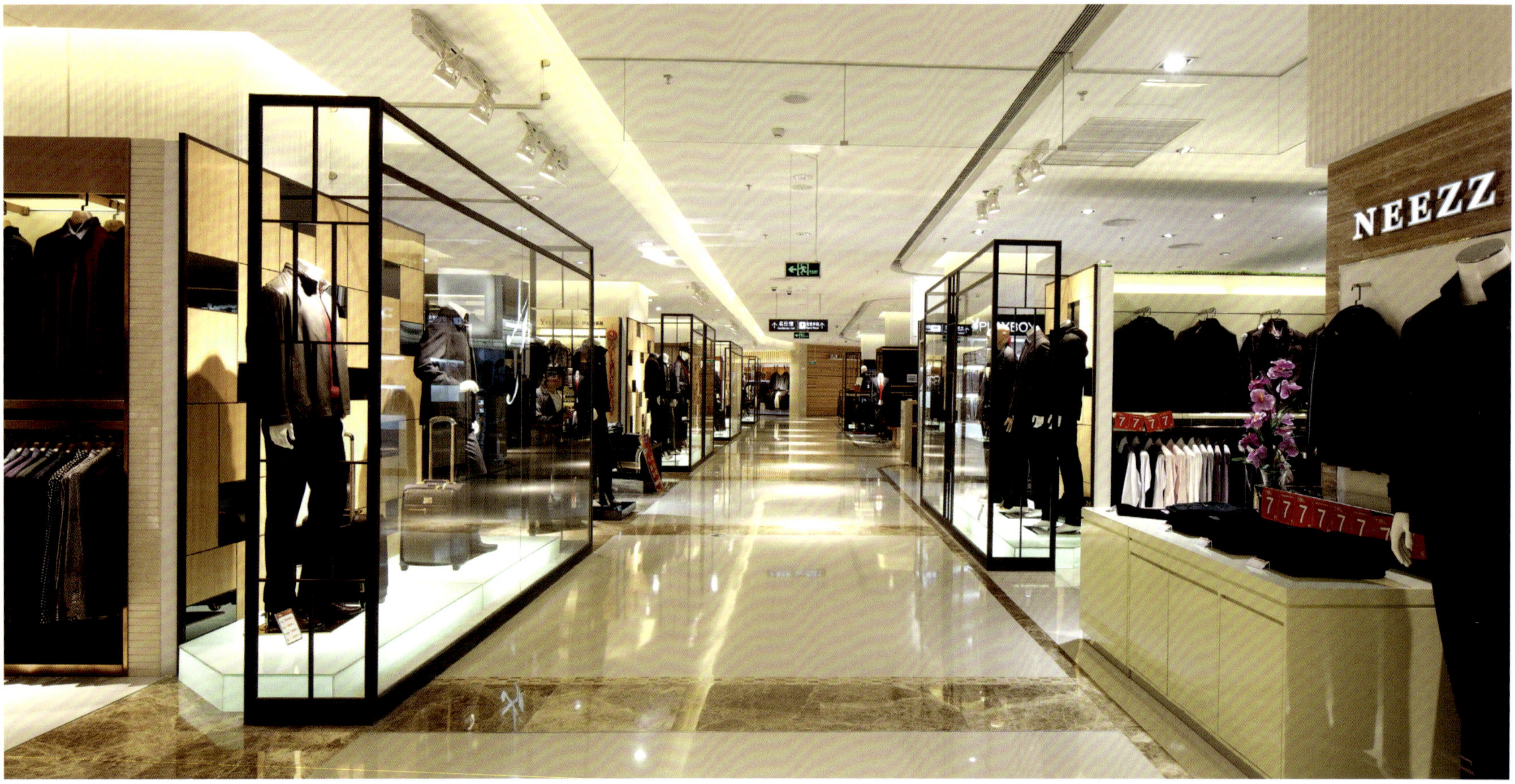
NEEZZ

# Facade Shopping Centre Limbecker Platz, Essen

## 埃森市林贝克广场购物中心门面

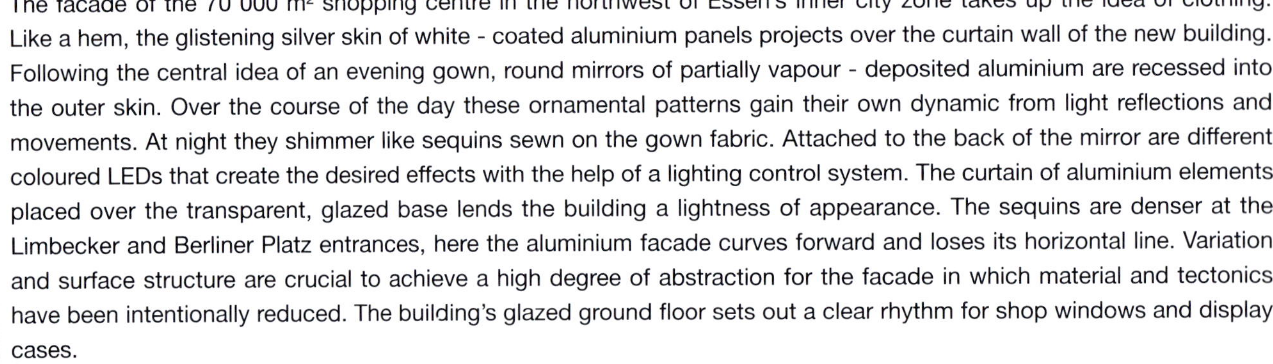

The facade of the 70 000 m² shopping centre in the northwest of Essen's inner city zone takes up the idea of clothing. Like a hem, the glistening silver skin of white - coated aluminium panels projects over the curtain wall of the new building. Following the central idea of an evening gown, round mirrors of partially vapour - deposited aluminium are recessed into the outer skin. Over the course of the day these ornamental patterns gain their own dynamic from light reflections and movements. At night they shimmer like sequins sewn on the gown fabric. Attached to the back of the mirror are different coloured LEDs that create the desired effects with the help of a lighting control system. The curtain of aluminium elements placed over the transparent, glazed base lends the building a lightness of appearance. The sequins are denser at the Limbecker and Berliner Platz entrances, here the aluminium facade curves forward and loses its horizontal line. Variation and surface structure are crucial to achieve a high degree of abstraction for the facade in which material and tectonics have been intentionally reduced. The building's glazed ground floor sets out a clear rhythm for shop windows and display cases.

**Location**
Limbecker Platz, 45127
Essen

**Area**
15,500 sq.m.

**Main materials**
Aluminium, glass, etc.

**Company**
Henn Architekten

**Designers**
Henn Architekten

**Photographer**
HG Esch

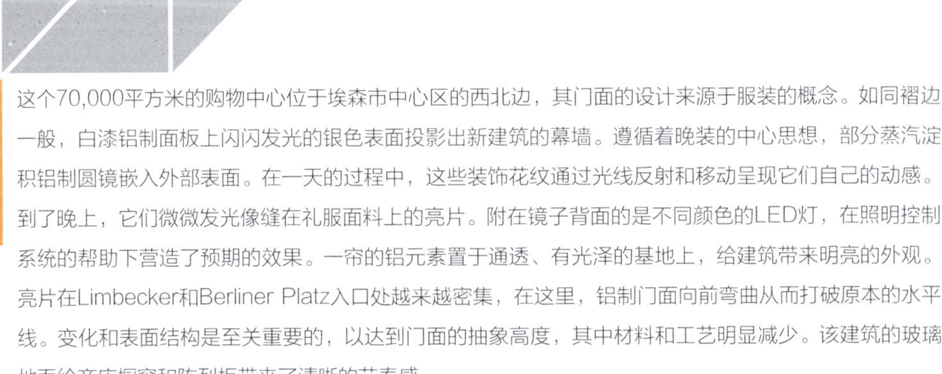

这个70,000平方米的购物中心位于埃森市中心区的西北边，其门面的设计来源于服装的概念。如同褶边一般，白漆铝制面板上闪闪发光的银色表面投影出新建筑的幕墙。遵循着晚装的中心思想，部分蒸汽淀积铝制圆镜嵌入外部表面。在一天的过程中，这些装饰花纹通过光线反射和移动呈现它们自己的动感。到了晚上，它们微微发光像缝在礼服面料上的亮片。附在镜子背面的是不同颜色的LED灯，在照明控制系统的帮助下营造了预期的效果。一帘的铝元素置于通透、有光泽的基地上，给建筑带来明亮的外观。亮片在Limbecker和Berliner Platz入口处越来越密集，在这里，铝制门面向前弯曲从而打破原本的水平线。变化和表面结构是至关重要的，以达到门面的抽象高度，其中材料和工艺明显减少。该建筑的玻璃地面给商店橱窗和陈列柜带来了清晰的节奏感。

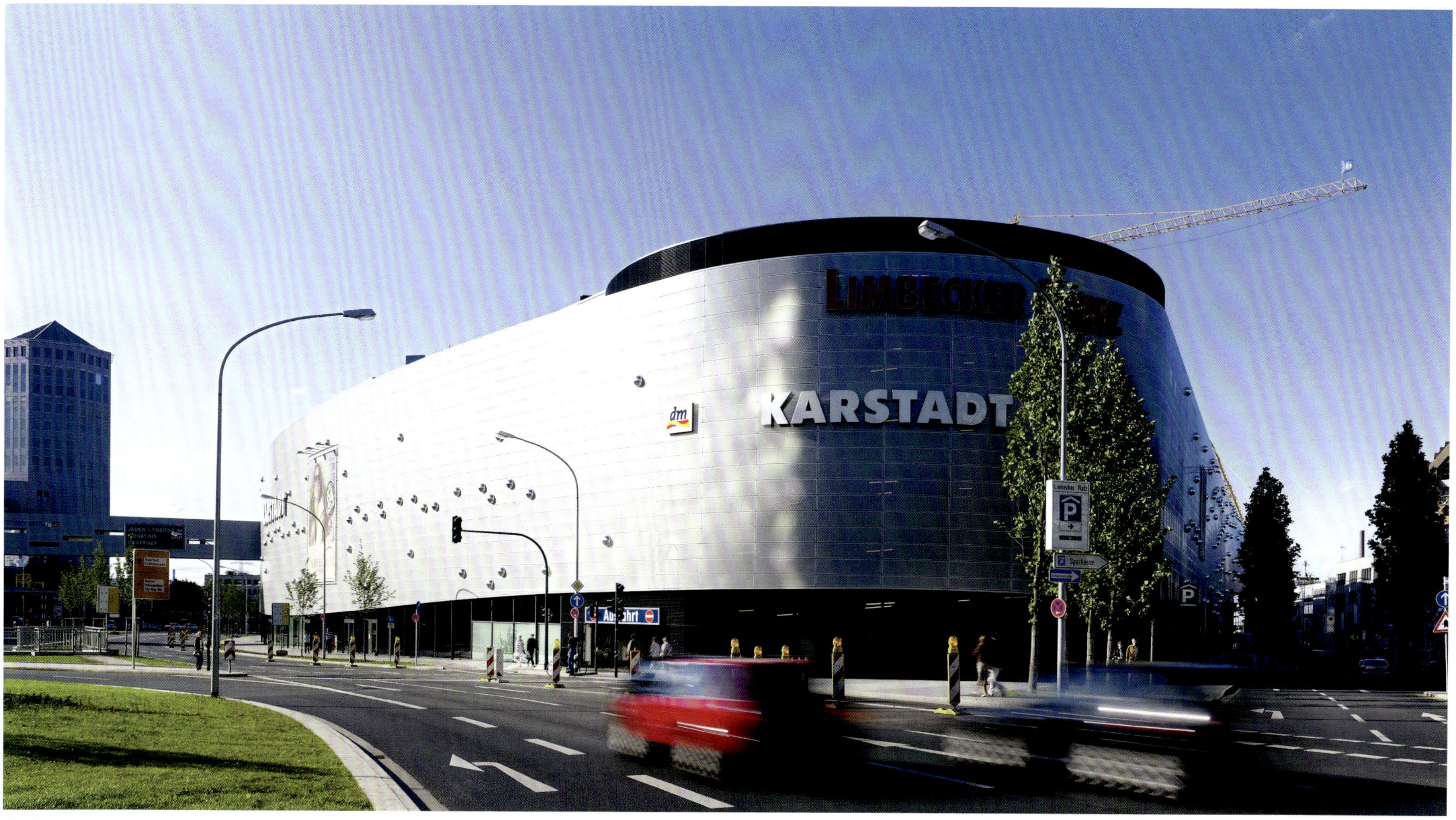

LIMB
KARSTADT
KARST
ESPRIT
dm

LIMBECKER PLATZ
APOTHEKE
ESPRIT
s.Oliver
KARSTADT
GameStop

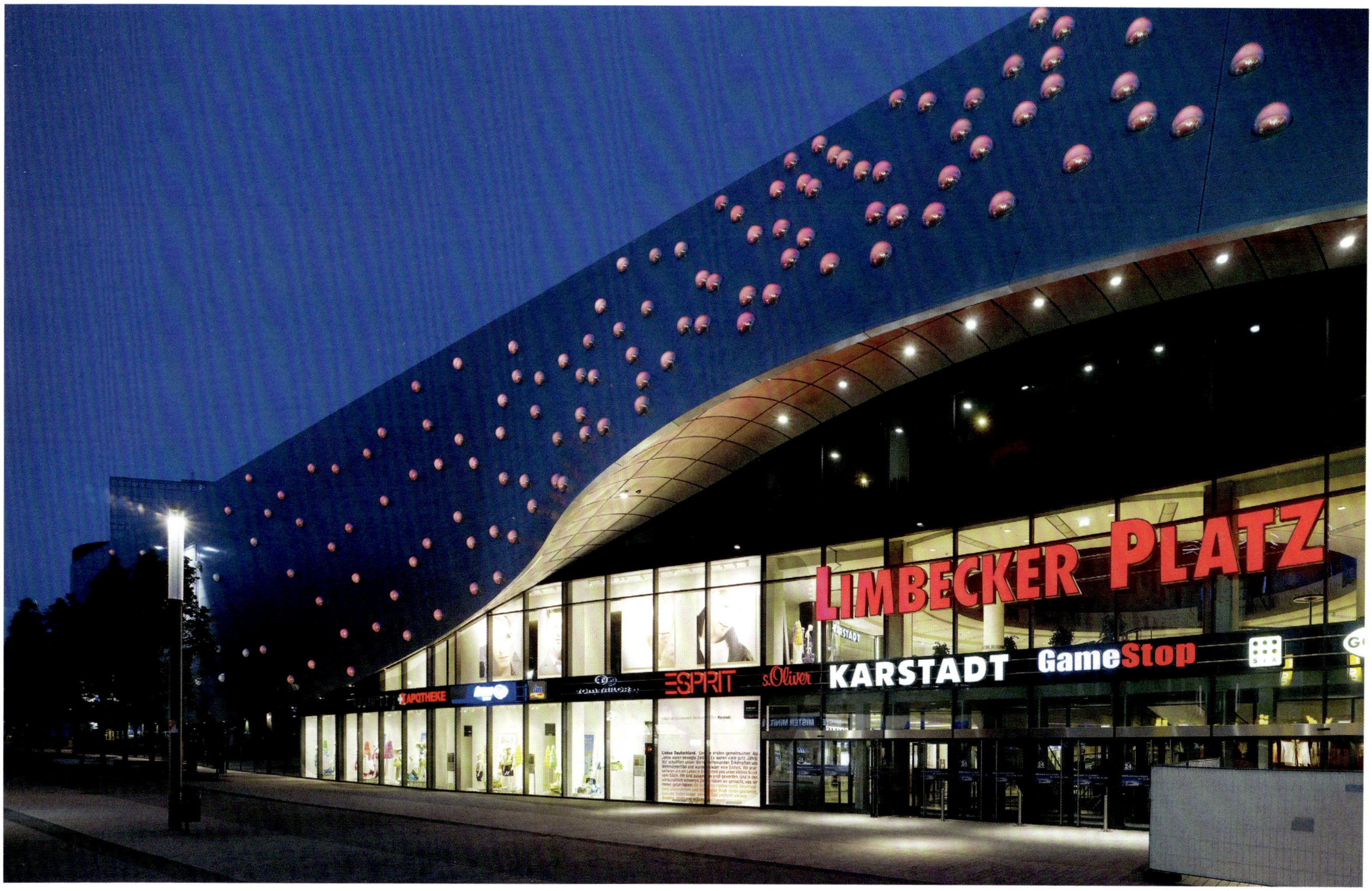

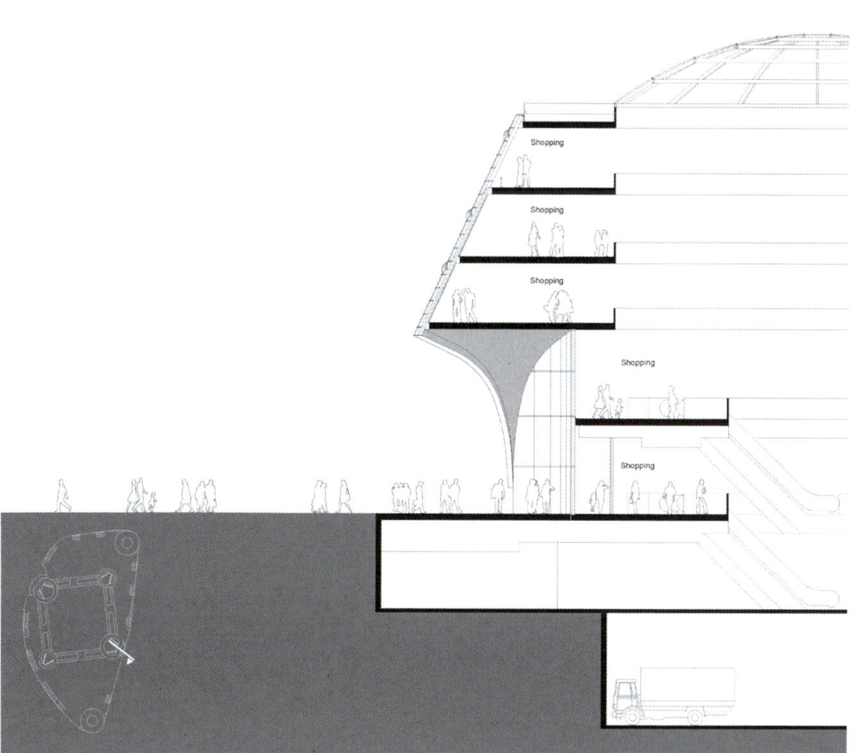

Shopping
Shopping
Shopping
Shopping
Shopping

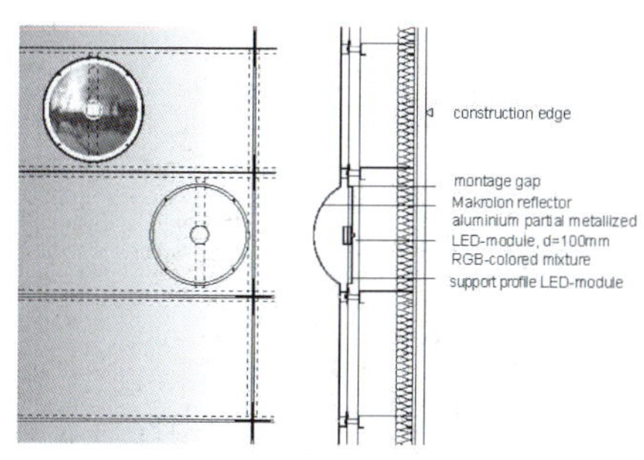

construction edge
montage gap
Makrolon reflector
aluminium partial metallized
LED-module, d=100mm
RGB-colored mixture
support profile LED-module

# Lentia Sky
# 伦蒂亚天空

Lentia City is a shopping center in the center of Linz with an expanded sales floor of 20,000 m². The main entrance, which was designed by LOVE, is situated on the main road - the busiest street in this part of the city.

The goal of this design was to first attract potential Lentia City visitors with a friendly and distinctive appearance in the street and then usher them into the interior of the shopping center. This goal was accomplished using the "Lentia Sky", a large silver panel fitted with spotlights that evoke a starry sky. It arches from the interior of the center into the street space and blends with the façade. Large, round façade openings with outward - curving plastic dome lights pierce the panel and provide natural lighting for the offices on the upper floors.

The Lentia City building is surrounded by 4-5 story old buildings. From an urban planning perspective, its façade, which flows into the center, and the receded entrance area create the feeling of an expansion of the existing street space, an effect which is supported by the bright, slightly reflecting metal façade.

The entrance area on the ground floor was designed in a conical shape which opens up into the street space like a funnel, thereby reinforcing the effect of drawing people in. On one side, a parade of shops leads from the mall to the street, while on the other side a panoramic elevator connects the five underground parking levels with the upper floors and also houses the fire escape.

**Location**
Hauptstraße 54 , 404
Linz, Austria

**Area**
750 sq.m.

**Main materials**
Aluminum, etc.

**Company**
LOVE architecture

**Photographer**
Jasmin Schuller

伦蒂亚城市中心是林茨市中心的一个购物中心，带有一个20,000平方米的扩充销售楼层。由LOVE建筑公司设计的主大门位于这个城市最繁忙的主要街道上。

设计目的是，首先在街上以一种友好的、独特的外观吸引伦蒂亚城市中心的潜在游客，然后将他们迎入购物中心的室内空间。这个目标通过利用刻有“伦蒂亚天空”的一块银色大面板而实现，面板上安装了射向星空的聚光灯。面板从中心内部往街区以拱形呈现，并与门面融为一体。大而圆的外墙开口带有向外弯曲的塑料顶灯，并在面板上形成穿孔，为楼上的办公区提供了自然光照。

伦蒂亚城市大厦的四周都是四到五层的旧楼。从城市规划的角度来看，流入中心的门面与流出的入口区营造了现有街道空间的扩充感，这是由明亮的、轻微反光的金属外观带来的效果。

一楼的入口区设计成圆锥形，像一个漏斗开辟到街道空间，因此强化了商场的吸引力。一方面，店铺的游行队伍从商场引向街道，而另一方面，全景电梯连接了地下五层停车场和地上楼层，并设有安全出口。

HOTEL
er Adler RESTAURANT

HOTEL
RESTAURANT
NEUROTH

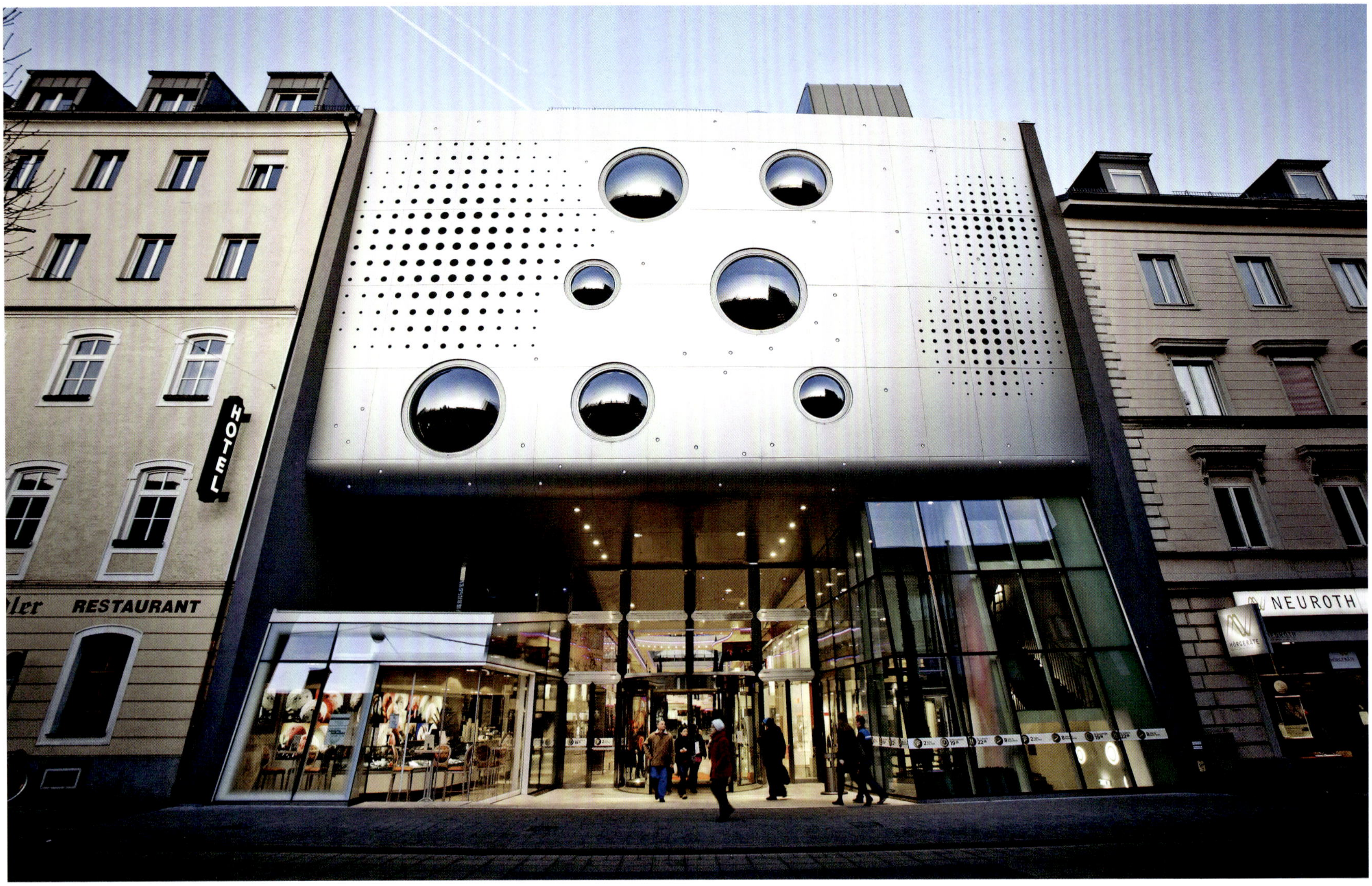

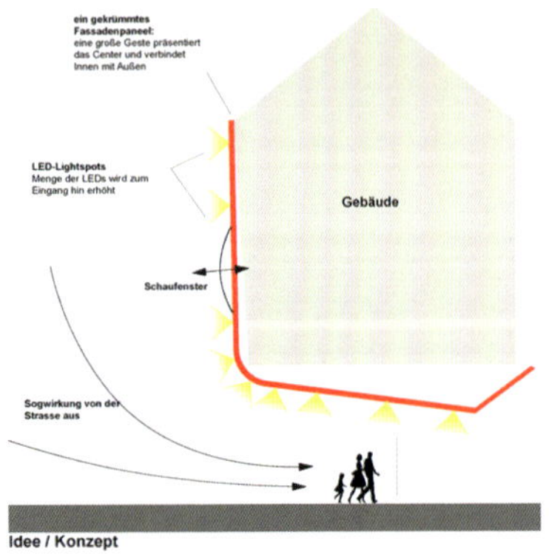

ein gekrümmtes
Fassadenpaneel:
eine große Geste präsentiert
das Center und verbindet
Innen mit Außen
LED-Lightspots
Menge der LEDs wird zum
Eingang hin erhöht
Gebäude
Schaufenster
Sogwirkung von der
Strasse aus

Idee / Konzept

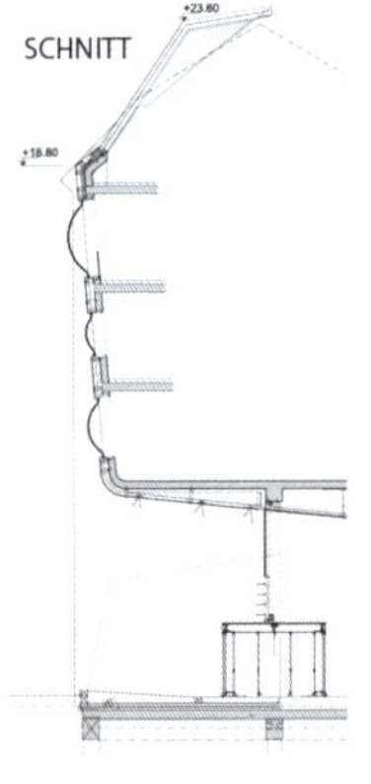

SCHNITT

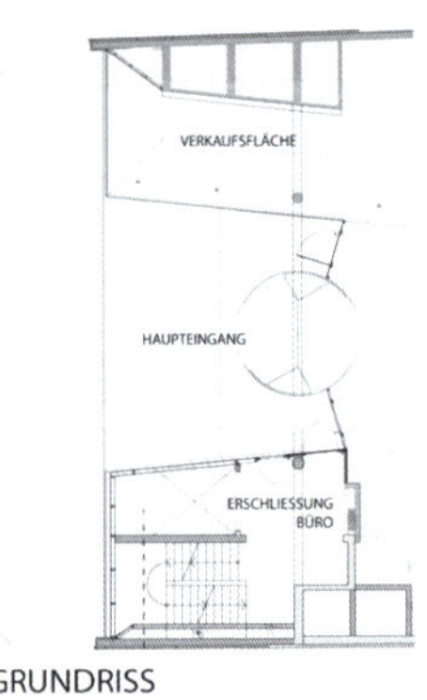

VERKAUFSFLÄCHE
HAUPTEINGANG
ERSCHLIESSUNG
BÜRO

GRUNDRISS

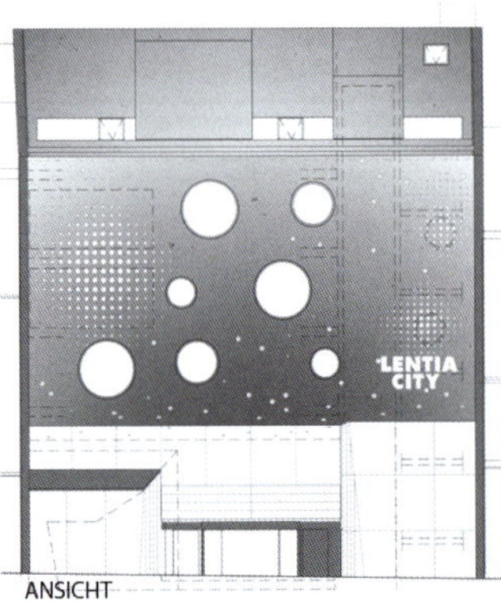

LENTIA
CITY

ANSICHT

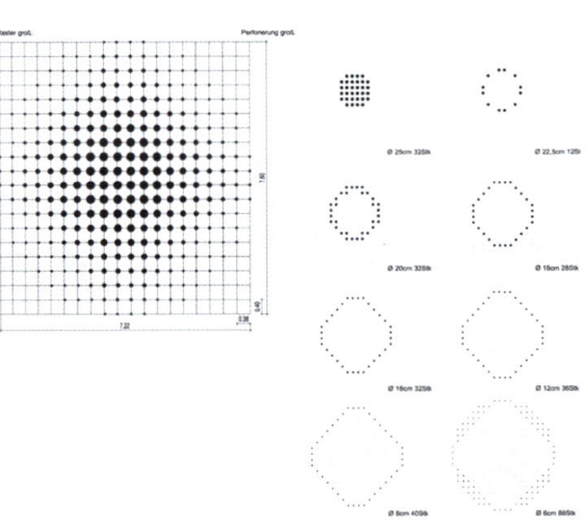

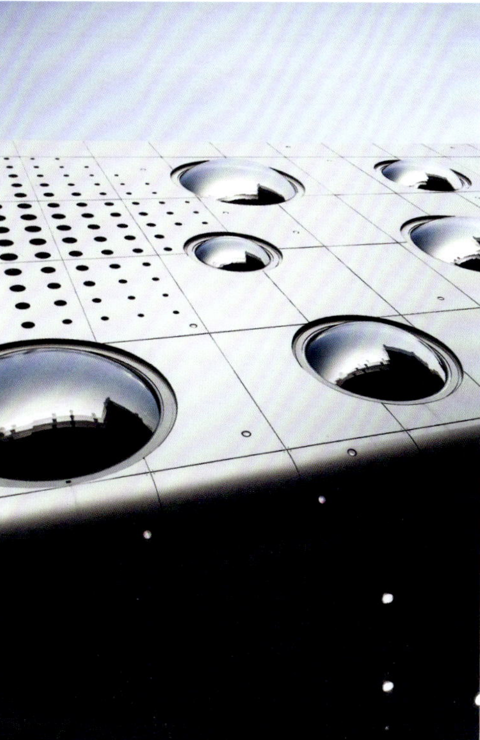

# Bahcelievler CarrefourSA Shopping Center

## Bahcelievler CarrefourSA购物中心

The building Bahcelievler CarrefourSA is composed of a hypermarket, a shopping mall and an enclosed parking. Located in European side of Istanbul, the project is adjacent to a busy highway, near to the Istanbul Ataturk Airport.

The conceptual design of the building takes the chaotic expression of the surrounding neighborhood as reference and brings it to a level of a "clear" architecture. It is separated from the sharp and visually crowded urban context with its calm curvilinear forms. The white streamline shape is also considered as a reference to the planes that land and take off every minute just across the site. Warm materials and intensive use of natural light offers visitors a serene escape from the city.

Bahcelievler CarrefourSA建筑由1个大型超市、1个购物商场和1个封闭式停车场组成。该项目坐落在伊斯坦布尔的欧洲区域，邻近繁忙的高速公路和伊斯坦布尔阿塔图尔克机场。

该建筑的设计概念借鉴了周围环境的混乱表达，给该建筑带来一个“清晰”的结构层。沉稳的曲线形式将该建筑与尖锐的且在视觉上拥堵的城市环境分离开来。白色的流线造型也参考了穿梭该地方的飞机在降落和起飞时每一分钟的形态。暖色的材料和自然光的集约使用为游客提供了一幅远离城市喧嚣的宁静画面。

**Location**
ISTANBUL, TURKEY

**Area**
65,000 sq.m.

**Company**
ERA PLANNING ARCHITECTURE CONSULTING CO. LTD.

**Designer**
ERA PLANNING ARCHITECTURE CONSULTING CO. LTD

TEKNOSA
Yemek alanı
ve açık hava teras
en üst katta!
TURKCELL

WC

four Axess'le
RA İNDİRİM

BURGER ING
baydöner
ISKENDER